Endocrinological Aspects of Alcoholism

Progress in Biochemical Pharmacology

Vol. 18

Series Editor
R. Paoletti, Milan

S. Karger · Basel · München · Paris · London · New York · Sydney

4th Annual Conference on Alcoholism, El Paso, Tex., Feb. 22–23, 1980

Endocrinological Aspects of Alcoholism

Volume Editors
F.S. Messiha and
G.S. Tyner, Lubbock, Tex.

65 figures and 27 tables, 1981

S. Karger · Basel · München · Paris · London · New York · Sydney

Progress in Biochemical Pharmacology

Vol. 16: Endogenous Peptides and Centrally Acting Drugs.
  24th Annual OHOLO Biological Conference, Zichron Ya'acov 1979.
  Levy, A. and Heldmann, E., Ness Ziona; Vogel, Z., Rehovot;
  Gutman, Y., Jerusalem (eds.)
  XVI + 160 p., 52 fig., 25 tab., 1980. ISBN 3-8055-0831-X

Vol. 17: Hormone and the Kidney
  6th Kanematsu Conference on the Kidney, Sydney 1980.
  Stokes, G.S. and Mahony, J.F., Sydney (eds.)
  VIII + 268 p., 64 fig., 20 tab., 1980. ISBN 3-8055-1090-X

National Library of Medicine, Cataloging in Publication
  Conference on Alcoholism (4th, 1980, El Paso, Tex.)
    Endocrinological aspects of alcoholism: 4th annual Conference on Alcoholism, El Paso, Tex., Feb. 22–23 1980
  Volume editors, F.S. Messiha and G.S. Tyner. – Basel, New York, Karger, 1981.
    (Progress in biochemical pharmacology; v. 18)
    Series of conferences initiated by Texas Tech University Health Sciences Center, School of Medicine, and cosponsored by Texas Research Institute on Mental Sciences and other agencies.
    1. Alcoholism – complications – congresses 2. Endocrine Diseases – etiology – congresses 3. Endocrine Glands – drug effects – congresses I. Texas Research Institute of Mental Sciences II. Texas Tech University. School of Medicine III. Messiha, Fathy S. IV. Tyner, G.S. V. Title VI. Series
    W1 PR666H v.18/WM 274 C7435 1980e

ISBN 3-8055-2689-X

Drug Dosage
  The authors and the publisher have exerted every effort to ensure that drug selection and dosage set forth in this text are in accord with current recommendations and practice at the time of publication. However, in view of ongoing research, changes in government regulations, and the constant flow of information relating to drug therapy and drug reactions, the reader is urged to check the package insert for each drug for any change in indications and dosage and for added warnings and precautions. This is particularly important when the recommended agent is a new and/or infrequently employed drug.

All rights reserved
  No part of this publication may be translated into other languages, reproduced or utilized in any form or by any means, eletronic or mechanical, including photocopying, recording, microcopying, or by any information storage and retrieval system, without permission in writing from the publisher.

© Copyright 1981 by S. Karger AG, P.O. Box, CH-4009 Basel (Switzerland)
Printed in Switzerland by Schüler AG, Biel
ISBN 3-8055-2689-X

Contents

Contributors VII
Preface XI

Neuroendocrinology and Neurophysiology

Rose, J.C.; Strandhoy, J.W., and *Meis, P.J.* (Winston-Salem, N.C.): Acute and Chronic Effects of Maternal Ethanol Administration on the Ovine Maternal-Fetal Unit 1
Greene, L.W. and *Hollander, C.S.* (New York, N.Y.): Alcohol and the Hypothalamus 15
Van Thiel, D.H. (Pittsburgh, Pa.): Hypothalamic-Pituitary-Gonadal Function in Liver Disease 24
Rawat, A.K. (Toledo, Ohio): Neuroendocrinological Implications of Alcoholism 35
Redmond, G.P. (Burlington, Vt.): Effect of Ethanol on Spontaneous and Stimulated Growth Hormone Secretion 58
Heine, M.W. (Lubbock, Tex.): Alcoholism and Reproduction 75

Fetal Alcohol Syndrome

Thadani, P.V. (Washington, D.C.): Fetal Alcohol Syndrome: Neurochemical and Endocrinological Abnormalities 83
Taylor, A.N.; Branch, B.J., and *Kokka, N.* (Los Angeles, Calif.): Neuroendocrine Effects of Fetal Alcohol Exposure 99
Abel, E.L. (Buffalo, N.Y.): Prenatal Effects of Beverage Alcohol on Fetal Growth 111
Lin, G.W.-J. (New Brunswick, N.J.): Fetal Malnutrition: A Possible Cause of the Fetal Alcohol Syndrome 115
Varma, S.K. and *Sharma, B.B.* (Lubbock, Tex.): Fetal Alcohol Syndrome 122

Biochemical Pharmacology

Zeiner, A. R. and Kegg, P. S. (Oklahoma City, Okla.): Effects of Sex Steroids on Ethanol Pharmacokinetics and Autonomic Reactivity 130

Dalterio, S.; Bartke, A.; Blum, K., and Sweeney, C. (San Antonio, Tex.): Marihuana and Alcohol: Perinatal Effects on Development of Male Reproductive Functions in Mice 143

Messiha, F. S. (Lubbock, Tex.): Subcellular Fractionation of Alcohol and Aldehyde Dehydrogenase in the Rat Testicles 155

Messiha, F. S. (Lubbock, Tex.): Epididymal Aldehyde Dehydrogenase: A Pharmacologic Profile 167

Behavioral Pharmacology

Tumbleson, M. E.; Dexter, J. D., and Middleton, C. C. (Columbia, Mo.): Voluntary Ethanol Consumption by Female Offspring from Alcoholic and Control Sinclair(S-1) Miniature Dams 179

Tumbleson, M. E.; Dexter, J. D., and Van Cleve, P. (Columbia, Mo.): Voluntary Ethanol Consumption, as a Function of Estrus, in Adult Sinclair(S-1) Miniature Sows 190

Pasley, J. N. and Powell, E. W. (Little Rock, Ark.): Effects of Hippocampal Lesions on Ethanol Intake in Mice 196

Messiha, F. S. (Lubbock, Tex.): Steroidal Actions and Voluntary Drinking of Ethanol by Male and Female Rats 205

Amino Acids

Fisher, S. E.; Atkinson, M.; Holzman, I.; David, R., and Van Thiel, D. H. (Pittsburgh, Pa.): Effect of Ethanol upon Placental Uptake of Amino Acids 216

Tan, A. T.; Dular, R., and Innes, I. R. (Montreal): Alcohol Feeding Alters (^{3}H)-Dopamine Uptake into Rat Cortical and Brain Stem Synaptosomes 224

Subject Index 231

Contributors

Abel, Ernest L., State of New York, Division of Alcohol and Alcohol Abuse, Research Institute on Alcoholism, Buffalo, N.Y. and State University of New York at Buffalo, N.Y. (USA)

Bartke, A., Department of Obstetrics and Gynecology, The University of Texas Health Sciences Center at San Antonio, San Antonio, Tex. (USA)

Blum, K., Division of Substance and Alcohol Misuse, Department of Pharmacology. The University of Texas Health Sciences Center at San Antonio, San Antonio, Tex. (USA)

Dalterio, Susan., Department of Pharmacology and Department of Obstetrics and Gynecology, The University of Texas Health Sciences Center at San Antonio, San Antonio, Tex. (USA)

Dullar, R., Department of Pharmacology and Therapeutics, Faculty of Medicine, University of Manitoba, Winnipeg (Canada)

Dexter, James D., Department of Neurology, University of Missouri Medical Center, School of Medicine and Sinclair Comparative Medicine Research Farm, Columbia, Mo. (USA)

Fisher, Stanley E., Pediatric Gastroenterology, Children's Hospital of Pittsburgh and the University of Pittsburgh School of Medicine, Pittsburgh, Pa. (USA)

Greene, Loren Wissner. Endocrine Division, Department of Medicine, Special Clinical Unit, New York University Medical Center, School of Medicine, New York, N.Y. (USA)

Hollander, Charles S., Endocrine Division, Department of Medicine, Special Clinical Unit, New York University Medical Center, School of Medicine, New York, N.Y. (USA)

Heine, M. Wayne, Department of Obstetrics and Gynecology, Texas Tech University Health Sciences Center, School of Medicine, Lubbock, Tex. (USA)

Innes, Ian R., Department of Pharmacology and Therapeutics, Faculty of Medicine, University of Manitoba, Winnipeg (Canada)

Lin, Grace W.-J., Center of Alcohol Studies, Rutgers University, New Brunswick, N.J. (USA)

Messiha, Fathy S., Departments of Pathology and Psychiatry, Director: Division of Toxicology, Department of Pathology, Texas Tech University Health Sciences Center, School of Medicine, Lubbock, Tex. (USA)

Pasley, James N., Department of Physiology, University of Arkansas, Little Rock, Ark. (USA)

Powell, Ervin W., Department of Anatomy, University of Arkansas, Little Rock, Ark. (USA)

Rawat, Arun K., Department of Biochemistry, Director: Alcohol Research Center, University of Toledo, Toledo, Ohio (USA)

Redmond, Geoffrey P., Departments of Pharmacology and Pediatrics, School of Medicine, University of Vermont, Burlington, Vt. (USA)

Rose, James C., Department of Physiology and Pharmacology, Bowman Gray School of Medicine, Winston-Salem, N.C. (USA)

Sharma, Bharat B., Department of Pediatrics, Texas Tech University Health Sciences Center, School of Medicine, Lubbock, Tex. (USA)

Strandhoy, J. W., Department of Obstetrics and Gynecology, Bowman Gray School of Medicine, Winston-Salem, N.C. (USA)

Thadani, Pushpa V., Alcohol Research, Veterans Administration Medical Center, Washington, D.C. (USA)

Tan, Ah-Ti., Department of Pharmacology and Therapeutics, University of Montreal and Centre de Recherche, Hôpital Louis-H. Lafontaine, Montreal (Canada)

Taylor, Anna Newman, Department of Anatomy, School of Medicine, University of California, Los Angeles, Center for the Health Sciences, Los Angeles, Calif. (USA)

Tumbleson, Myron E., College of Veterinary Medicine, School of Medicine and Sinclair Comparative Medicine Research Farm, University of Missouri, Columbia, Mo. (USA)

Varma, Surendra K., Associate Chairman, Department of Pediatrics, Texas Tech University Health Sciences Center, School of Medicine, Lubbock, Tex. (USA)

Van Thiel, David H., Gastroenterology, Department of Medicine, University of Pittsburgh School of Medicine, Pittsburgh, Pa. (USA)

Zeiner, Arthur R., Department of Psychiatry and Behavioral Sciences, The University of Oklahoma, Health Sciences Center, School of Medicine, Oklahoma City, Okla. (USA)

Preface

The present volume represents the proceedings of the Fourth Annual Conference on Alcoholism held in El Paso, Texas, during February, 1980. These series of conferences have been initiated by Texas Tech University Health Sciences Center, School of Medicine, and cosponsored by Texas Research Institute of Mental Sciences, Baylor College of Medicine, Texas Commission on Alcoholism and the Committee on Physician Health and Rehabilitation. Contributors to the proceedings came from conference participants and from invited investigators.

Endocrine abnormalities in chronic alcoholism has been known for decades and endocrine research in alcoholism made enormous strides in recent years due to the development of more specific techniques for the measurements of certain hormones. However, an understanding of the pathophysiology, psychopathogenesis and the underlying mechanisms is far from complete. Moreover, information related to the subject has been widely dispersed throughout the literature and is only known to the expert.

This monograph, as its title states, evaluates existing concepts, reviews the latest research findings, surveys the clinical data base and provides a framework in which clinical and experimental findings can be reviewed and discussed by anatomists, biochemists, endocrinologists, gynecologists, neurologists, pediatricians, pharmacologists, physiologists and behavioral scientists.

We hope that the present proceedings will present a comprehensive interdisciplinary survey of recent approaches to the study of alcohol-mediated endocrine dysfunction.

F.S. Messiha and *G.S. Tyner*, Lubbock, Tex.

Neuroendocrinology and Neurophysiology

Acute and Chronic Effects of Maternal Ethanol Administration on the Ovine Maternal-Fetal Unit[1]

J.C. Rose[a], J.W. Strandhoy[a], P.J. Meis[b]

[a] Department of Physiology and Pharmacology and [b] Department of Obstetrics/Gynecology, Bowman Gray School of Medicine, Winston-Salem, N.C., USA

Introduction

Ethanol is a small, freely diffusable molecule which readily crosses the placenta [4, 10, 15, 17]. Since ethanol crosses the placenta, it is capable of affecting the developing fetus, and considerable evidence in the literature suggests that this is the case. Currently, the effects of maternal ethanol consumption during pregnancy on the developing neonate are of intense interest in both the scientific and lay communities. In particular, the reports of *Jones* et al. [12] and *Jones and Smith* [13] have stimulated interest in this area. These workers noted a pattern of multiple congenital abnormalities associated with chronic maternal alcohol abuse during pregnancy. They termed this pattern of congenital anomalies the 'fetal alcohol syndrome'. The initial discrimination of the fetal alcohol syndrome has been confirmed by other investigators [7, 18] although the existence of a specific pattern of anomalies has been questioned [19, 20]. There is also some debate as to whether alcohol is the sole cause of the anomalies noted, or if other factors such as cigarette smoking or malnutrition may play some role. However, it is clear that the excessive consumption of alcohol during pregnancy represents a significant risk to normal intra- and extrauterine development. In particular, mental retardation and growth deficiency have been the two most common problems noted in children of women who abuse alcohol during pregnancy [3]. Studies in several animal species have also shown deleterious

---

[1] Supported in part by NCARA grants 7704 and 7804 and by grant HD 11210-01 from the National Institutes of Health.

effects of excessive alcohol consumption during pregnancy. Dismorphogenesis and growth retardation as well as decreased litter size, increased incidence of infant mortality, and decreased incidence of live births have been associated with excessive alcohol consumption during pregnancy [14, 22, 25]. In addition, there are biochemical defects noted in various organ systems of newborns exposed to alcohol *in utero* [5, 24]. Such studies all suggest that alcohol is a teratogenic agent which produces profound alterations in several organ systems in the fetus and neonate.

Because orderly changes in fetal endocrine systems are necessary for normal perinatal growth and development, we have begun to examine some of the effects of alcohol on both fetal endocrine function and fetal growth. The model chosen for study is the chronically cannulated fetal sheep *in utero*. With this model it is possible to study the fetus in a relatively undisturbed intrauterine environment in the absence of complications caused by anesthesia, and in the absence of nonspecific changes inherent in studies on the acutely exteriorized fetus.

Material and Methods

Time-dated pregnant ewes are brought to the laboratory at day 100 of gestation and assigned to either control or alcohol-treated groups in a random fashion. A catheter is placed into the jugular vein pericutaneously and henceforth animals received either alcohol infusions ($2 \text{ g} \cdot \text{kg}^{-1}$ body weight) or isocaloric dextrose infusions daily. Control and experimental animals are pair-fed in an attempt to maintain equivalent nutritional status in both groups. At approximately 120 days of pregnancy, fetal cannulae are implanted under strict aseptic conditions as described in a previous publication [21]. After a 3-day recovery period from surgery, blood samples are obtained from the ewe resting in her cage and the fetus for the determination of various parameters. The animals are followed until birth occurs or for as long as the preparation is viable.

All blood samples were obtained in lightly heparinized syringes, immediately transferred to chilled centrifuge tubes, and the plasma separated by centrifugation at $1,000 g$ for 10 min at 4 °C. Plasma was then aspirated, aliquoted, and stored frozen for subsequent analysis of hormones and/or alcohol levels. All blood pressures were measured with P23 DB pressure transducers (Stathem Instruments Division, Gould Inc., Oxnard, Calif.). Heart rates were monitored with a cardiotachometer triggered from the arterial pressure signal. All pressure and heart rate recordings were made on a Hewlett-Packard recorder.

Thyroid activity was tested by injecting thyrotropin-releasing hormone (TRH) intravenously into the fetuses of control and alcohol-treated ewes. All hormone measurements were done with radioimmunoassay (RIA). Thyroxine (T_4) and triiodothyronine (T_3) were measured with reagents supplied by Radioassay Systems Laboratories, Carson, Calif. T_4

cross-reacts in the T_3 assay less than 1%, and T_3 cross-reacts in the T_4 assay less than 1%. Cortisol was measured according to the method of *Abraham* et al. [1] using an antiserum that is highly specific for cortisol. Plasma alcohol levels were measured with alcohol dehydrogenase and NAD by the method of *Bernt and Gutmann* [2].

Results

Kinetics and Cardiovascular Effects of Ethanol

Figures 1 and 2 show the maternal and fetal plasma alcohol concentrations before, during, and after an alcohol infusion (2 g·kg^{-1} body weight) to the pregnant ewe. The peak blood alcohol concentration in the pregnant ewe was noted at the end of the 2-hour infusion (240±6 mg·dl^{-1}, n = 7). In the fetus plasma ethanol levels also peaked at 2 h (190±9 mg·dl^{-1}, n = 7), and the levels were significantly lower (p<0.05) than those observed in the pregnant ewe. The maternal rate of elimination following the ethanol infusion was 40 mg·dl^{-1} per hour, while the fetal rate of elimination was 10 mg·dl^{-1} per hour. There was a highly significant correlation in plasma alcohol concentrations in samples obtained simultaneously from mother and baby during and after the maternal ethanol infusion (fig. 3).

The alcohol infusions produced a significant increase in maternal systolic and diastolic pressures, 79±3 (mean ± SEM) to 88±4 Torr and 55±1 to 63±3 Torr, respectively, after 1 hour (fig. 4). At the end of 2 h the tendency for the increases was present, but the differences were no longer statistically significant. Mean arterial pressure was also significantly elevated at the end of 1 h, 64±2 to 73±3 Torr (p<0.05). Maternal heart rate was increased at 1 and 2 h after the ethanol infusion was initiated. Infusions of dextrose produced no changes in maternal blood pressure or heart rate. Fetal blood pressure and heart rate were not affected by maternal infusions of alcohol or dextrose (n = 4).

Ethanol Effects on Endocrine Function

Maternal and fetal plasma cortisol levels before and after a 2-hour infusion of ethanol into the mother are shown in figures 5 and 6. The ethanol infusion was associated with a significant increase (p<0.01) in both maternal and fetal plasma cortisol concentrations. Maternal plasma cortisol levels increased from 13.4±2 to 39.4±9 and fetal plasma cortisol levels increased from 36.9±18 to 101.4±57.6. Infusions of dextrose produced no increase in fetal or maternal plasma cortisol levels (n = 3).

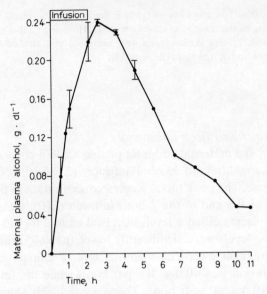

Fig. 1. Maternal plasma alcohol concentrations before, during, and after a 2-hour infusion of alcohol (2 g·kg^{-1} body weight). Points and brackets represent the mean ± SEM, with n = 7. Single points represent 2 animals.

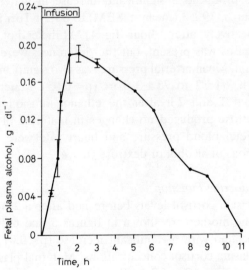

Fig. 2. Fetal plasma alcohol concentrations before, during, and after a 2-hour infusion of alcohol to the ewes depicted in figure 1. Points and brackets represent the mean ± SEM, with n = 5–7. Single points represent 2 animals.

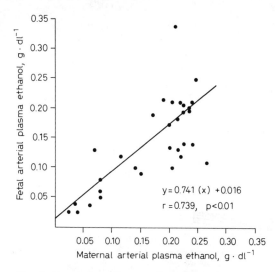

Fig. 3. Correlation of maternal and fetal plasma ethanol concentrations during and after maternal ethanol infusion of 2 g·kg^{-1} body weight administered over a 2-hour period.

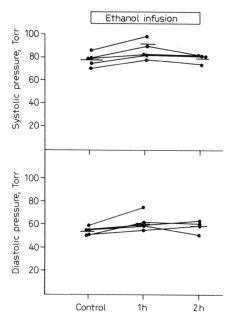

Fig. 4. Maternal systolic and diastolic blood pressure before and after infusion of alcohol (2 g·kg^{-1} body weight over 2 h). Changes in both pressures were significant at 1 h.

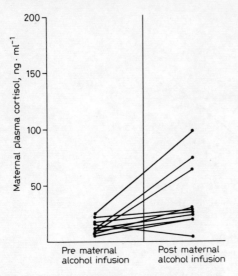

Fig. 5. Maternal plasma cortisol levels at the beginning and end of a 2-hour infusion of alcohol (2 g·kg^{-1} body weight).

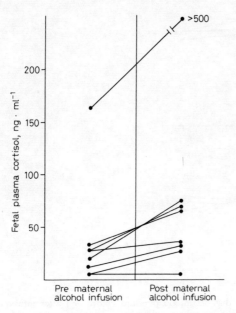

Fig. 6. Fetal plasma cortisol concentration at the beginning and end of a 2-hour infusion of alcohol (2 g·kg^{-1} body weight) to the pregnant ewe.

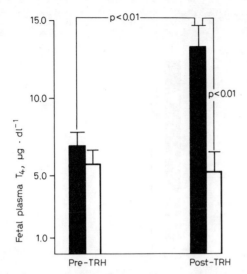

Fig. 7 Changes in plasma T$_4$ concentrations following a 250-µg injection of TRH into fetuses of control (■, n = 9) and alcohol-treated ewes (□, n = 5). Bars and brackets represent the mean ± SEM. The average gestational age for the control animals was 131 ± 1 day and that of the alcohol treated group was 135 ± 2 days.

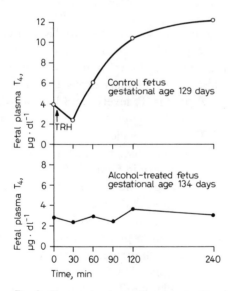

Fig. 8. Changes in plasma T$_4$ concentrations in a single control fetus and in a single fetus exposed to alcohol. The TRH (250 µg) was given to both animals where indicated by the arrow in the top portion of the figure.

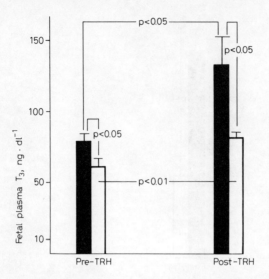

Fig. 9. Changes in plasma T_3 concentration following a 250-μg injection of TRH into fetuses of control (■, n = 9) and alcohol-treated ewes (□, n = 5). Bars and brackets represent the mean ± SEM. The average gestational age for the control animals was 131 ± 1 day and that of the alcohol treated group was 135 ± 2 days.

In figure 7 are shown the fetal plasma T_4 responses to TRH in control animals (n = 9) and animals exposed to alcohol *in utero* (n = 5). Administration of TRH increased plasma T_4 from 6.9 ± 0.8 to 12 ± 1.3 μg · dl^{-1} in the control animals. In the animals exposed to alcohol *in utero,* TRH did not increase fetal plasma T_4 levels – 5.8 ± 0.9 μg · dl^{-1} pre-TRH, 5.1 ± 0.5 μg · dl^{-1} post-TRH. In figure 8 the plasma T_4 responses to a 250-μg injection of TRH in a single control and a single alcohol-treated fetus are shown. The T_3 responses to TRH in these same control and alcohol-treated animals are shown in figure 9. Basal T_3 levels were significantly different in the two groups. The resting plasma T_3 concentration in the control animals was 78.9 ± 4.8 ng · dl^{-1} and in the alcohol-treated animals it was 60.8 ± 5.9 ng · dl^{-1}. In both groups TRH produced a statistically significant elevation in plasma T_3 concentration. However, the control animals showed a much greater increase ($p < 0.05$) of plasma T_3 levels than the alcohol-treated animals did – 125.6 ± 21.8 mg · dl^{-1} ves. 73.2 ± 4.8 ng · dl^{-1}. Plasma T_3 responses to TRH (250 μg) in a single control and a single alcohol-treated animal are shown in figure 10.

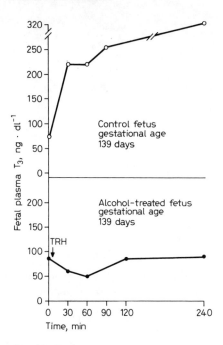

Fig. 10. Changes in plasma T_3 concentration in a single control fetus and in a single fetus exposed to alcohol. The TRH (250 μg) was given to both animals where indicated by the arrow in the bottom portion of the figure.

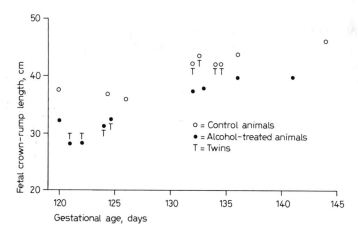

Fig. 11. Effect of maternal alcohol treatment (2 g·kg^{-1} body weight) on fetal crown-rump length. The pregnant ewes received alcohol infusions daily beginning on day 100 of pregnancy.

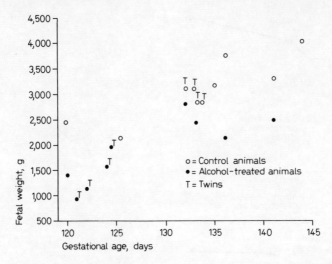

Fig. 12. Effect of maternal alcohol treatment (2 g·kg^{-1} body weight) on fetal weight. The pregnant ewes received alcohol infusions daily beginning on day 100 of pregnancy.

Effects of Chronic Ethanol Exposure on Fetal Body Length and Body Weight

Fetal crown-rump lengths in control animals and animals exposed to alcohol *in utero* are shown in figure 11. When the animals are age matched, the alcohol-treated animals have crown-rump lengths less than the controls, and two-way analysis of variance (for age and treatment) indicates this difference is significant ($p < 0.05$). Fetal weight comparisons are shown in figure 12. When the animals are age matched, the alcohol-treated animals show lower body weights than the controls and two-way analysis of variance (for age and treatment) indicates this difference is also significant ($p < 0.05$). The mean duration of alcohol exposure *in utero* was 25 ± 1 day in the alcohol-treated animals.

Discussion

It is obvious from this study and from the work of others that alcohol readily crosses the placenta and rapidly appears in the fetal blood. Our preliminary data suggests that fetal plasma alcohol concentrations lag somewhat behind maternal plasma alcohol concentrations during the period of infusion. However, there is a close correlation between the plasma alcohol concentration observed in the fetus and that observed in the mother. Disappearance of alcohol from the maternal circulation appears to occur more rapidly than does the disappearance of alcohol from the fetal circulation. This is consistent with observations made in human fetuses and newborns where alcohol elimination occurs much more slowly than it does in adults [9, 23, 26]. Whether or not this difference in the disappearance rate represents a difference in metabolism by the fetus is yet to be established.

Our results indicate that ethanol produces small but significant increases in maternal systolic, diastolic, and mean arterial blood pressure and in maternal heart rate. It may be that these changes are mediated via an increase in circulating catecholamine levels. In our chronic animal preparations, alcohol did not change fetal heart rate or blood pressure. Others have reported that alcohol changes fetal blood pressure and heart rate [8, 15]. However, their work was in acutely prepared preparations and thus subject to complications induced by the presence of anesthesia and stress. Both of these factors can produce marked changes in fetal hemodynamics.

Our findings that alcohol infusions produce increases in both fetal and maternal plasma cortisol levels are consistent with reports that ethanol is a stimulus to steroid secretion in men and in experimental animals [6, 11, 16]. Presumably, the increases in fetal and maternal plasma cortisol levels following an alcohol infusion are secondary to increases in fetal and maternal plasma ACTH levels. It is doubtful that the increases in fetal plasma cortisol are the result of diffusion of cortisol from the mother to the fetus since the placenta represents a significant barrier for cortisol in this species.

Our data indicate that chronic ethanol exposure *in utero* alters fetal thyroid function. The T_4 responses to TRH were abolished in fetuses exposed to alcohol, and the T_3 responses to TRH were attenuated markedly in the same animals. In addition, the basal levels of T_3 were reduced in the alcohol-treated animals. At present, it is unclear if this reduction in responsiveness to TRH is due to a lack of a fetal TSH response to the tripeptide, or to a defect at the level of the fetal thyroid with it unresponsive to TSH. It is also possible that alcohol may increase the metabolic clearance of TSH and/or the thyroid hormones. More experiments must be done to establish the cause of this lack of responsiveness in the alcohol-treated animals.

In the present experiments, alcohol exposure during the last third of gestation resulted in fetuses with crown-rump lengths less than control animals and with body weights less than age-matched control animals. This suggests that alcohol exposure during this period of pregnancy inhibits fetal growth in this sheep model. It is recognized, however, that to firmly establish this point more animals must be studied to expand the gestational age range of the controls as well as the experimental animals, and to obtain enough data to develop regression lines describing growth rates for both the control and treatment groups. In summary, our results indicate that alcohol exposure *in utero* alters fetal adrenal function and fetal thyroid function. It may be that these alterations in endocrine status in the fetus play a role in the apparent growth retardation found in the animals studied to date.

Summary

The purpose of this study was to examine, in the pregnant ewe and its fetus, some of the physiological consequences of acute and chronic ethanol exposure. Ethanol was infused intravenously (2 g/kg/day over 2 h) to pregnant ewes from day 100 of pregnancy to term. Control animals received isocaloric infusions of 5% dextrose. Animals were pair-fed and allowed water *ad lib*. Maternal (n = 5) systolic, diastolic, and mean blood pressures and heart rate rose significantly by 1 h after starting ethanol, whereas fetal (n = 4) blood pressure and heart rate did not change during ethanol infusion. Maternal ethanol infusion produced a significant rise ($p < 0.01$) in both fetal (n = 8) and maternal (n = 10) plasma cortisol levels. Peak blood ethanol concentration was significantly higher in the ewe (240 ± 6 mg·dl^{-1}, n = 7) than in the fetus (190 ± 9 mg·dl^{-1}, n = 6) at the end of the 2-hour infusion. Maternal rate of elimination after ethanol infusion was terminated was 40 mg·dl^{-1} per hour, while fetal elimination was 10 mg·dl^{-1} per hour. Body weights and crown-rump lengths of fetuses from 0.82 to 1.0 gestation were significantly less in ethanol-treated animals than in age-matched control animals. Fetal plasma thyroxine and triiodothyronine increments following thyrotropin-releasing hormone administration were significantly less in alcohol-treated than in control animals. Thus, chronic exposure to ethanol during the latter part of gestation impaired fetal growth and altered fetal endocrine function in these animals.

Acknowledgment

The authors wish to express their appreciation for technical assistance provided by Messrs. *Harvey Boose* and *Willie Hunt* and by Ms. *Helen Edwards*, Ms. *Linda McDonald*, and Ms. *Susan Bellinger*. Also, the authors wish to thank Ms. *Stephanie Burgoyne* for typing the manuscript.

References

1 Abraham, G.E.; Buster, J.E.; Teller, R.C.: Radioimmunoassay of plasma cortisol. Analyt. Lett. *5:* 757–759 (1972).
2 Bernt, E.; Gutmann, I.: Ethanol determination with alcohol dehydrogenase and NAD. Meth. enzym. Analysis *3:* 1499–1502 (1974).
3 Clarren, S.K.; Smith, D.W.: The fetal alcohol syndrome. New Engl. J. Med. *298:* 1063–1067 (1978).
4 Dilts, P.V.: Placental transfer of ethanol. Am J. Obstet. Gynec. *107:* 1195–1198 (1970).
5 Elis, J.; et al.: The effect of alcohol administration during pregnancy on concentration of noradrenaline, dopamine and 5-hydroxtryptamine in the brain of offspring of mice. Acta nerv. super, Praha *18:* 220–221 (1976).
6 Ellis, F.W.: Adrenal cortical function in experimental alcoholism in dogs. Proc. Soc. exp. Biol. Med. *120:* 740–744 (1965).
7 Hall, B.D.; Orenstein, W.A.: Noonan's phenotype in an offspring of an alcoholic mother. Lancet *i:* 680–681 (1974).
8 Horiguchi, T.; et al.: Effect of ethanol upon uterine activity and fetal acid-base state of the rhesus monkey. Am J. Obstet. Gynec. *109:* 910–917 (1971).
9 Idanpaan-Heikkila, J.E.; et al.: Placental transfer of ^{14}C-ethanol. Am. J. Obstet Gynec. *110:* 426–431 (1971).
10 Idanpaan-Heikkila, J.; et al.: Elimination and metabolic effects of ethanol in mother, fetus, and newborn infant. Am. J. Obstet. Gynec. *112:* 387–393 (1972).
11 Jenkins, J.S.; Connolly, J.: Adrenocortical response to ethanol in man. Br. med. J. *ii:* 804–805 (1968).
12 Jones, K.L.; et al.: Pattern of malformation in offspring of chronic alcohol mothers. Lancet *i:* 1267–1271 (1973).
13 Jones, K.L.; Smith, D.W.: Recognition of the fetal alcohol syndrome in early infancy. Lance *ii:* 999–1001 (1973).
14 Kronick, J.B.: Teratogenic effects of ethyl alcohol administered to pregnant mice. Am. J. Obstet. Gynec. *124:* 678–680 (1976).
15 Mann, L.I.; et al.: Placental transport of alcohol and its effect on maternal and fetal acid-base balance. Am. J. Obstet. Gynec. *122:* 837–844 (1975).
16 Merry, J.; Marks, V.: Plasma-hydrocortisone response to ethanol in chronic alcoholics. Lancet *i:* 921–923 (1969).
17 Nicloux, M.: Sur le passage de l'alcool ingéré de la mère au fœtus, en particulier chez la femme. Cr. Séanc. Soc. Biol. *51:* 980–982 (1899).
18 Palmer, R.H.; et al.: Congenital malformations in offspring of a chronic alcoholic mother. Pediatrics, Springfield *53:* 490–494 (1974).
19 Ouellette, E.M.; Rosett, H.L.: A pilot prospective study of the fetal alcohol syndrome at the Boston City Hospital. II. The infants. Ann. N.Y. Acad. Sci. *273:* 123, 129 (1976).
20 Ouellette, E.M.; et al.: Adverse effects on offspring of maternal alcohol abuse during pregnancy. New Engl. J. Med. *297:* 528–530 (1977).
21 Rose, J.C.; et al.: Developmental aspects of the pituitary-adrenal axis response to hemorrhagic stress in lamb fetuses *in utero.* J. clin. Invest. *61:* 424–432 (1978).

22 Sandor, S.: The influence of ethyl alcohol on the developing chick embryo. Rev. Roum. Embryol. *5:* 167–171 (1968).
23 Seppala, M.; et al.: Ethanol elimination in a mother and her premature twins. Lancet *i:* 1188–1189 (1971).
24 Thadani, P.V.; et al.: Effects of maternal ethanol ingestion on amine uptake into synaptosomes of fetal and neonatal rat brain. J. Pharmac. exp. Ther. *200:* 292–297 (1977).
25 Tze, W.G.; Lee, M.: Adverse effects of maternal alcohol consumption on pregnancy and fetal growth in rats. Nature, Lond. *257:* 479–480 (1975).
26 Wagner, L.; et al.: Effect of alcohol on premature newborn infants. Am. J. Obstet. Gynec. *108:* 308–315 (1970).

J.C. Rose, PhD, Department of Physiology and Pharmacology,
Bowman Gray School of Medicine, Winston-Salem, NC 27103 (USA)

Alcohol and the Hypothalamus

Loren Wissner Greene, Charles S. Hollander[1]

Endocrine Division, New York University School of Medicine, New York, N.Y., USA

Our knowledge of the hypothalamus, much like the larger field of neurobiology, is advancing at a bewildering pace with simultaneous contributions from many diverse disciplines. It is perhaps not surprising therefore that our basic fund of knowledge of the effects of alcohol on the endocrine system in general and the hypothalamic-pituitary axis in particular is also expanding rapidly. As in neurobiology, however, the ultimate answers are not known; and at this stage, it is awkward to conceptualize too firmly from any isolated findings. In the following review which seeks to put our knowledge in the appropriate perspective, the effects of alcohol on the hypothalamus (hormone secretion, behavior and basic physiologic responses of the organism) are described successively under separate headings; although, as will be apparent, there is considerable overlap in many instances.

The clinical manifestations of hypothalamic dysfunction include: sexual abnormalities including hypogonadism, diabetes insipidis and hypernatremia; psychic disturbances such as rage and hallucinations; feeding disorders ranging from hyperphagia and obesity to anorexia and

---

[1] We wish to thank Ms. *Jean Price* for her help and assistance with typing and editing this manuscript.

emaciation; somnolence and sleep rhythm reversal; problems of thermoregulation [26]; and dysfunction of the autonomic nervous system with sphincter disturbances and cardiac arrhythmias [23]. Although all of these findings are prevalent in chronic alcoholics, this by no means proves that alcoholics have primary hypothalamic disease.

Direct confirmation that the locus of alcoholic action is the hypothalamus is currently impossible. For one thing, it is difficult to dissect out the effects of alcohol on stress, pain tolerance, nutritional status and vitamin levels, and liver and other chronic diseases from the direct and specific effects of alcohol on hormone secretion [32]. Acute and chronic alcohol consumption cause differing hormonal results. Also, the effects of alcohol on humans are not necessarily the same as in animals. At differing quantities of alcohol, the hypothalamic response may vary. In addition, the stress or anesthesia necessary for cannulation of the hypophyseal portal circulation will affect hormone release. Hypothalamic hormones are difficult to assay, especially in peripheral blood where they are present in trace amounts and may be rapidly degraded. In addition, there may be methodological problems in the specificity of radioimmunoassay because of cross-reactivity with molecules of similar structure or larger potentially inactive molecules containing identical amino acid sequences. It is therefore necessary to utilize *in vitro* hypothalamic systems to study the direct effects of alcohol [15].

Many of the behavioral effects of alcohol may relate to its actions on the hypothalamus. The different stages of alcohol dependency (alcohol preference [1], intoxication, tolerance, and withdrawal) may be due to a central or hypothalamic action [2]. Perception of peripheral hormone levels centrally may affect alcohol preference: hypothyroid rats have an increased intake [14]. Alcohol may release endogenous opioids that enhance lateral hypothalamic self-stimulation behavior in rats since this effect is blocked by naloxone [19]. An increase in endorphin secretion may also cause the desire to imbibe in humans.

Alcohol may acutely cause the release of central and hypothalamic norepinephrine. During the 'excitation phase' of alcohol intoxication, there is clinical evidence of sympathetic overactivity [2, 14]. Although much of this activity is due to peripheral catecholamine release, there may be some central output. The consequent depletion of norepinephrine may explain the inhibition or sleep phase, and the lack of hormone responsivity to alcohol-induced hypoglycemia [2, 12]. As tolerance develops, a compensatory increase in norepinephrine synthesis may have occurred. Therefore, during periods of withdrawal, when increased synthesis is maintained without increased release and breakdown, signs of sympathetic stimulation are again manifest. Alcohol can relieve these symptoms of withdrawal, possibly by again increasing norepinephrine depletion [14].

Other theories to explain the hyperexcitability of alcohol withdrawal include dopamine depletion [13], depletion of inhibitory amines such as GABA or an increase in false neurotransmitters [24].

Alcohol is also known to disturb many of the mammalian rhythms of neuroendocrine function, probably through its action on the hypothalamus. The normal maturational increase in LH secretion, the gonadotropin cyclic surges at ovulation, and the diurnal variation in such hormones as GH, prolactin, and ACTH have all been deranged. Sleep disturbances induced by alcohol include a suppression of REM sleep with decreased frequency, delayed onset and decreased depth of REM [13]. Some chronic alcoholics have hypersomnolence which may be due to increased uptake of aromatic amines into the central nervous system or to decreased cholinergic output [24]. Since sleep is important in the diurnal secretion of many hormones [14] it is easy to understand why alcohol can upset the physiological patterns of release. In the following paragraphs the effects of alcohol on hypothalamic hormones will be discussed. These hormones include those of known or putative structure that affect the posterior or anterior pituitary gland.

Hypothalamic Hormones Affecting the Posterior Pituitary

Alcohol acutely decreases the release of vasopressin and oxytocin from the posterior pituitary gland either through its effect on the hypothalamus or directly on the posterior pituitary [16]. The decreased secretion of oxytocin in response to alcohol has been used therapeutically in the treatment of premature labor [14]. The response of vasopressin to osmotic stimuli is decreased especially in young men. The ensuing diuresis can be overcome by exogenous vasopressin. Chronically, alcohol can cause hypernatremia due to vasopressin depletion [16]. This depletion of vasopressin may also account for other actions of alcohol, for example, improvement in memory has been noted after the use of vasopressin in two patients with Korsakoff's syndrome. Vasopressin therapy has also decreased the severity of alcohol-induced hypothermia [10].

Hypothalamic Hormones Affecting the Anterior Pituitary

Hypothalamic Hormones of Known Structure

Luteinizing Hormone-Releasing Hormone

The effects of alcohol on sexual behavior have been noted for centuries. As shown in the often cited reference from *Shakespeare's Macbeth* [27]: It provokes the desire but takes away the performance'. Although alcohol increases sexual arousal in man, possibly through a surge in luteinizing hormone (LH) due to decreased testosterone, it decreases sexual performance, in part through its CNS effects, causing impotence in alcoholics [25]. It has also been shown to interfere with copulatory behavior in male rats and dogs [14].

Biphasic effects of alcohol on gonadotropin secretion are seen: at low doses, there is an increased output of LH (along with signs of CNS stimulation such as increased neuronal firing and excitation); however, at doses greater than 1.5 g/kg, there is a decrease in LH (that may precede the decrease in testosterone temporally) along with evidence of CNS depression [7, 8].

Alcohol interferes with pulsatile LH secretion like other sleep-related events [5, 14]. It prevents the normal LH increase at maturation in male rats [8, 28, 32]. The alcohol-induced inhibition of ovulation (and the associated rise in LH) in female rats and mice can be reversed by giving LH, therby showing that the site of interference is pituitary or higher [4]. Possibilities include decreased synthesis or release of LH, decreased effect of LHRH on the pituitary, or decreased synthesis of or release of luteinizing hormone-releasing hormone (LHRH) from the hypothalamus. Since there is a normal response to LHRH in most studies in man and rodents after acute alcohol intake and there may be an exaggerated response to LHRH in some chronic alcoholics with elevated basal LH and FSH, the locus of alcohol action may be the hypothalamus [32]. The normal increase in LH produced by castration is prevented by alcohol [31]. This blockade is also reversed by LHRH again suggesting a central site of inhibition. The decreased response to clomiphene [9] noted in chronic alcoholic men may be due to a pituitary, hypothalamic, or higher effect of alcohol [30].

There have been many hypotheses to explain the change in gonadotropin secretion. Other agents, besides alcohol, that decrease catecholamine content or increase serotonin levels in the CNS have been shown to decrease LH secretion. The disturbed sleep patterns of alcoholics may affect hormone secretion. In other states associated with decreased appetite, decreased food intake, and weight loss, there is decreased LH [28]. The hyperestrogenized state of chronic alcoholics with liver disease may prevent a compensatory rise in LH and FSH when testosterone is low [30].

In summary, numerous studies have documented effects of alcohol on gonadotropin secretion and many have shown that one locus of alcohol action is central. Preliminary direct observations from our laboratory using an *in vitro* system suggests that this would be at the hypothalamic level [unpublished data].

Thyrotropin-Releasing Hormone

An increase in prolactin with an increased response of prolactin to thyrotropin-releasing hormone (TRH) 4 h after acute alcohol ingestion in normal subjects has been found [32]. During withdrawal the combination of increased basal prolactin with decreased response of prolactin and TSH to TRH and increased GH found in some chronic alcoholics has been attributed to a central increase in dopaminergic activity [20, 33]. Weeks after alcohol intake ceases, the blunted TRH response remains detectable in some individuals. This excess dopamine may have restitutional effects since dopamine agonists such as apomorphine HCl, *L*-dopa and amphetamines alleviate symptoms of alcohol withdrawal [20].

Of course, the increased dopaminergic activity and blunted TRH test may be unrelated. Depression, which is common in alcoholics, and stress with increased cortisol levels are well-known causes of blunted TRH tests. Also, it is unclear whether or not hypothalamic dopamine levels are affected by alcohol [3].

SRIF (Somatostatin)

Acutely, there is an increase in GH secretion (which may occur along with an increase in ACTH secretion) in healthy individuals given 1.5 ml/kg ethanol [14]. Alcohol alters the circadian rhythm of GH secretion: it delays the normal rise in GH with onset of sleep [12]. In addition, it blunts the reponse of GH release to normal stimuli such as hypoglycemia [6], arginine, vasopressin, and propranol and glucagon in chronic alcoholics and during withdrawal [6, 12]. Conceivably these effects are mediated through central control by increased SRIF, decreased GRF or a direct effect on the pituitary. In our laboratory we have demonstrated that alcohol increases SRIF secretion in organ culture of rat hypothalamus [unpublished data]. Thus SRIF may act as a master inhibitory hormone, causing inhibition of GH, TRH release, and TRH effect on TSH release. We have demonstrated the decrease in TRH release from physiological concentrations of SRIF in dispersed cells of rat hypothalamus [17]. It is interesting to note that the decreased GH response to arginine occurs in acromegalics [29] as well as normal persons given alcohol [22]. The decreased response of GH to hypoglycemia may be an isolated phenomenon or may occur in conjunction with decreased responses of other hormones such as prolactin, ACTH and sympathetic hormones [21, 32].

Hypothalamic Hormones of Putative Structure

Corticotropin-Releasing Factor

There must be an effect of alcohol on the pituitary not mediated by the hypothalamus since some patients with hypothalamic lesions show increased cortisol levels when they are given alcohol acutely. Again, this does not prove that there is no hypothalamic control when the hypothalamus is intact. After an acute intraperitoneal injection of alcohol, adrenal steroid production increases in rats due to release of pituitary ACTH. This increase in corticosterone (the major rat glucocorticoid) is blocked by morphine, so again there may be hypothalamic control [14].

In addition, although both increases and decreases in ACTH secretion have been observed in chronic alcoholics, this may relate to different loci of alcohol action. For example, mice lose the circadian rhythm of ACTH secretion when given alcohol chronically [11]. Also, in many chronic alcoholics a 'pseudo-Cushing's syndrome' is seen. These patients resemble patients with Cushing's disease in that they show an abnormally high threshold for suppression of ACTH release by exogenous and endogenous glucocorticoids and lack the normal diurnal variation of ACTH secretion. They may also hyper-respond to stimuli for ACTH secretion such as metyrapone and lysine vasopression [18, 32]. It is unclear whether this alcoholic syndrome is due to increased release of a hypothalamic factor (i.e. corticotropin-releasing factor) or if a direct effect of alcohol on the pituitary release of ACTH is more important. In addition, such factors as stress may be more important than the direct effect of alcohol and alcoholic withdrawal on the hypothalamic-pituitary axis. However, stress is not the only reason for the steroid hormone release noted in alcohol withdrawal: diazepam or barbiturates can block the stress and symptoms of alcohol withdrawal without affecting the release of cortisol. Alcohol itself can decrease cortisol and prevent withdrawal symptoms [22, 32].

In other chronic alcoholics, decreased ACTH secretion is found either selectively (isolated ACTH deficiency) or in conjunction with a decreased response of other hormones such as GH, and the catecholamines to hypoglycemia [32].

In conclusion, alcohol may act on the hypothalamus to cause diverse behavioral and neuroendocrine syndromes. Admittedly, it is difficult to provide a unified explanation for all these effects at the present time. The increased availability of sensitive and specific assays for hypothalamic hormones and suitable paradigms for assessing release *in vitro* should provide the researcher and clinician with the tools for elucidating these mechanisms in the future.

References

1 Amit, Z.; Levitan, D.E.; Meade, R.G.: The effects of pre-exposure to ethanol, ventral hypothalamic lesions and stimulation on the acquisition of ethanol preference. Commun. Psychopharmacol. *1:* 147–155 (1977).
2 Anokhina, I.P.: Neurochemical aspects in pathogenesis of alcohol and drug dependence. Drug Alcohol Depend. *4:* 265–273 (1979).
3 Bacopoulos, N.G.; Bhatnagar, R.K.; Orden, L.S. van: The effects of subhypnotic doses of ethanol on regional catecholamine turnover. J. Pharmac. exp. Ther. *204:* 1–10 (1978).
4 Blake, C.A.: Paradoxical effects of drugs acting on the central nervous system on the preovulatory release of pituitary luteinizing hormone in prooestrous rats. J. Endocr. *79:* 319–326 (1978).
5 Boyar, R.M.: Sleep-related endocrine rhythms; in Reichlin, Baldessarini, Martin, The hypothalamus, pp. 373–386 (Raven Press, New York 1978).
6 Chalmers, R.J.; Bennie, E.H.; Johnson, R.H.; Masterton, G.: Growth hormone, prolactin, and corticosteroid responses to insulin hypoglycaemia in alcoholics. Br. med. J. *i:* 745–748 (1978).
7 Cicero, T.; Badger, T.M.: A comparative analysis of the effects of narcotics, alcohol and the barbiturates on the hypothalamic-pituitary-gonadal axis. Adv. exp. med. Biol. *853:* 95–115 (1977).
8 Cicero, T.; Badger, T.M.: Effects of alcohol on the hypothalamic-pituitary-gonadal axis in the male rat. J. Pharmac. exp. Ther. *201:* 427–433 (1977).

9 Cicero, T.J.; Meyer, E.R.; Bell, R.D.: Effects of ethanol on the hypothalamic pituitary-luteining hormone axis and testicular steroidogenesis. J. Pharmac. exp. Ther. *208:* 210–215 (1979).
10 De Weid, D.: Hormonal influences on motivation, learning, memory and psychosis; in Krieger, Hughes, Neuroendocrinology, pp. 194–204 (Sinauer Associates, Sunderland 1980).
11 Ellis, F.W.: Effect of ethanol on plasma corticosterone levels. J. Pharmac. exp. Ther. *153:* 121–127 (1966).
12 Ganda, O.P.; Sawin, C.T.; Iber, F.; Glennon, J.A.; Mitchell, M.L.: Transient suppression of growth hormone secretion after chronic ethanol intake. Alcoholism *2:* 297–299 (1978).
13 Gitlow, S.E.; Dziedzic, S.W.; Dziedzic, L.B.: Persistent abnormalities in central nervous system function (long-term tolerance) after brief ethanol administration. Drug Alcohol Depend. *2:* 453–468 (1977).
14 Gordon, G.G.; Southren, A.L.: Metabolic effects of alcohol on the endocrine system; in Lieber, Metabolic aspects of alcoholism, pp. 249–302 (University Park Press, Baltimore 1977).
15 Greene, L.W.; Hollander, C.S.: Sex and alcohol. The effects of alcohol on the hypothalamic-pituitary-gonadal axis. Alcoholism *4:* 1–5 (1980).
16 Helderman, J.H.; Vestal, R.E.; Rowe, J.W.; Tobin, J.D.; Andres, R.; Robertson, G.L.: The response of arginine vasopressin to intravenous ethanol and hypertonic saline in man. The impact of aging. J. Geront. *33:* 39–47 (1978).
17 Hollander, C.S.; Greene, L.W.; Rosman, L.; Yamauchi, K.; Richardson, S.B.; Thaw, C.; D'Eletto, R.: Reciprocal local feedback of somatostatin (SRIF) and thyrotropin releasing hormone (TRH) in dispersed cell culture of rat hypothalamus (Abstract). Clin. Res. *28:* 479A (1980).
18 Lamberts, S.W.J.; De Jong, F.H.; Birkenhager, J.C.: Biochemical characteristics of alcohol-induced pseudo-Cushing's syndrome. J. Endocr. *80:* 62P–63P (1979).
19 Lorens, S.A.; Sainati, S.M.: Naloxone blocks the excitatory effect of ethanol and chlordiazepoxide on lateral hypothalamic self-stimulation behavior. Life Sci. *23:* 1359–1364 (1978).
20 Loosen, P.T.; Prange, A.J., Jr.; Wilson, I.C.: TRH (Protirelin) in depressed alcoholic men. Behavioral changes and endocrine responses. Aurchs. gen. Psychiat. *36:* 540–547 (1979).
21 Marks, V.: Alcohol-induced hypoglycaemia. Br. med. J. *i:* 238 (1978).
22 Marks, V.; Wright, J.W.: Endocrinological and metabolic effects of alcohol. Proc. R. Soc. Med. *70:* 337–344 (1977).
23 Martin, J.B.; Reichlin, S.; Brown, G.M.: Neurologic manifestations of hypothalamic disease. Clin. Neuroendocr. 247–273.
24 Noble, E.P.; Tewari, S.: Metabolic aspects of alcoholism in the brain; *in* Lieber, Metabolic aspects of alcoholism, pp. 149–185 (University Park Press, Baltimore 1977).
25 Mendelson, J.H.; Mello, N.K.: Biologic concomitants of alcoholism. New Engl. J. Med. *301:* 912–921 (1979).
26 Ritchie, J.M.: The aliphatic alcohols; in Goodman and Gilman The pharmacological basis of therapeutics; 3rd ed., pp. 143–158 (Macmillan, New York 1966).

27 Shakespeare, W.: Macbeth, Act II, Scene 3.
28 Symons, A.M.; Marks, V.: The effects of alcohol on weight gain and the hypothalamic-pituitary-gonadotrophin axis in the maturing male rat. Biochem. Pharmac. 24: 955–958 (1975).
29 Tamburrano, G.; Tamburrano, S.; Gambardella, S.; Andreani, D.: Effects of alcohol on growth hormone secretion in acromegaly. J. clin. Endocr. Metab. 42: 193–196 (1976).
30 Van Thiel, D.H.; Lester, R.; Sherins, R.J.: Hypogonadism in alcoholic liver disease. Evidence for a double defect. Gastroenterology 67: 1188–1199 (1974).
31 Van Thiel, D.H.; Gavaler, J.S.; Cobb, C.F.; Sherins, R.J.; Lester, R.: Alcohol-induced testicular atrophy in the adult male rat. Endocrinology 105: 888–895 (1979).
32 Wright, J.: Endocrine effects of alcohol. Clin. Endocr. Metab. 7: 351–367 (1978).
33 Ylikahri, R.H.; Huttunen, M.O.; Harkonen, M.; Leino, T.; Helenius, T.; Liewendahl, K.; Karonen, S.L.: Acute effects of alcohol on anterior pituitary secretion of the tropic hormones. J. clin. Endocr. Metab. 46: 715–720 (1978).

Dr. C.S. Hollander, Department of Medicine, Endocrine Division,
New York University Medical Center, School of Medicine, 550 First Avenue,
New York, NY 10016 (USA)

Hypothalamic-Pituitary-Gonadal Function in Liver Disease

David H. Van Thiel

Division of Gastroenterology at the University of Pittsburgh School of Medicine, Pittsburgh, Pa., USA

For years the testicular atrophy, gynecomastia, changes of body hair and vascular abnormalities seen in males with cirrhosis have been attributed to metabolic imbalance secondary to the liver disease [2, 16]. This assumption has been based upon the fact that the liver has a central role in sex steroid metabolism, detoxification and excretion [26]. Therefore, changes in sex steroid metabolism in such men were expected and were presumed to lead to the observed changes. Considerable evidence has accumulated recently, however, to suggest that despite histologically advanced liver disease, the metabolic clearance rate of sex steroids is not altered significantly [10, 19, 24]. Moreover, considerable evidence has accumulated to suggest that alcohol abuse, one of the major factors of liver disease in the Western World, can produce dysfunction of the hypothalamic-pituitary-gonadal axis even in the absence of liver disease [11, 28, 31, 35, 37]. Therefore, we have assessed hypothalamic-pituitary-gonadal function in groups of men with differing types of liver disease but with similar degrees of liver disease as determined by biochemical and histologic evaluation to determine whether the liver disease per se or the etiology of the liver disease determines the presence or the character of hypothalamic-pituitary-gonadal dysfunction, should such occur. The following is a summary of our findings:

Alcoholic Liver Disease (Table I)

The Male
Chronic alcoholic men are both hypogonadal and feminized [33]. Their hypogonadism is manifested by testicular atrophy, impotence and

Table I. Defects in the function of the hypothalamic-pituitary-gonodal axis present in individuals with alcoholic liver disease

A Gonadal defects
 1 Endocrine failure
  a Reduced sex steroid production and secretion (testosterone and progesterone in the male and female, respectively)
  b Reduced 17α-hydroxylase
  c Reduced 3β-hydroxysteroid dehydrogenase/Δ5,Δ4-isomerase
  d Altered redox state ↓ NAD/NADH ↑
 2 Reproductive failure
  a Testicular atrophy with loss of germ cells in the male
  b Absence of corpora lutea in the female

B Hypothalamic-pituitary defects
 1 Reduced basal gonadotropin secretion considering the degree of gonadal failure present
 2 Inadequate clomiphene responses
 3 Inadequate LH response to LRF

C Hepatic defects and defects in other tissues effecting hypothalamic-pituitary-gonadal function
 1 Enhanced levels of sex steroid binding globulin (effectively reducing free sex steroid levels)
 2 Enhanced aromatase activity and portosystemic shunting (increases androgen conversion to estrogen)
 3 Enhanced 5α-reductase activity
 4 Reduced levels of male specific hepatic estrogen binding protein
 5 Increased levels of hepatic estrogen receptor
 6 Reduced levels of androgen receptor

loss of libido, all of which occur in as many as 70–80% of such individuals [20]. Feminization in alcoholic men is manifested by the presence of a female escutcheon and palmer erythema in as many as 50%, the presence of spider angiomata in 40% and gynecomastia in as many as 20% [28].

Because the seminiferous tubules comprise 90–95% of the testicular volume, the gonadal atrophy present in chronic alcoholic men reflects primarily a reduction in the mass of the seminiferous tubules, the reproductive compartment of the testes. Thus, histologic studies of the testes obtained from chronic alcoholics demonstrate a reduction in the cross-sectional area of the seminiferous tubules as well as a marked reduction in the number and kinds of the germ cells they contain [3, 18, 20]. With severe and/or prolonged alcohol abuse, all germ cells may be lost and only Sertoli cells can be found within the seminiferous tubules [28, 33].

Failure of the endocrine compartment of the testes of chronic alcoholic men also is well established, in addition to the obvious histologic changes seen in the reproductive compartment of the testes. Leydig cell injury is demonstrated by the markedly reduced plasma testosterone levels. Thus, 50% of the alcoholic men have testosterone levels below the lower limit of normal, while those who have normal levels generally have levels within the lower half of the normal range [13, 28]. The endocrine failure of the testes of chronic alcoholic men is even more apparent when the plasma-free testosterone (biologically active) level is determined [1, 33]. Presumably, as a consequence of associated liver disease, as well as Leydig cell failure, the plasma concentration of sex steroid binding globulin is markedly increased (on mean eightfold) in such men [28, 32]. Normally, this plasma protein binds 99–99.5% of the plasma testosterone [5]. When its level increases, the fraction of free biologically active testosterone is reduced for any given absolute level of plasma testosterone. Therefore, as a result of a marked increase in the sex steroid binding globulin concentration and a marked reduction in total plasma testosterone, both of which are seen in chronic alcoholic males, the free biologically active testosterone level is markedly reduced [21].

Several mechanisms have been suggested a being responsible for the reduced testosterone production seen in alcoholic men [29, 34]. First, the recent demonstration of an alcohol dehydrogenase in testicular tissue suggests that as a result of alcohol metabolism, a change in the testicular redox state may occur [30]. It is thus possible that within the testes NADH accumulates while NAD levels are reduced as a result of ethanol metabolism. Because NAD is required for normal $\Delta 5$, $\Delta 4$-isomerase and 3β-hydrosteroid reductase as well as 17β-reductase activity, the activities of these 3 key enzymes in steroidogenesis may be reduced within the testes of alcohol-ingesting men [29, 34]. Alternatively, or additionally, acetaldehyde may accumulate within the testes as a consequence of ethanol metabolism and thus act as a mitochondrial poison and limit testosterone production at the level of side chain cleavage [7]. Experimental evidence exists for each of these possibilities.

As is the case for the abnormalities seen in the endocrine compartment, several mechanisms have been suggested as possible etiological factors responsible for the reduction in reproductive tissue seen in the testes of chronic alcoholic men [29, 30, 34]. First, testosterone has been shown to be essential for normal spermatogenesis. Thus, it is at least possible that primary alcohol-induced Leydig cell injury may be responsible for the failure of the reproductive compartment of the testes. Interference with normal vitamin A metabolism also may explain, at least in part, the reproductive failure [30]. Specifically, it has been established that vitamin A is essential for normal spermatogenesis. The molecular form of vitamin A which is biologically active within the testes is retin*al*. Retin*ol* is delivered to the testes and converted to retin*al* by an alcohol dehydrogenase. Ethanol, however, has a fiftyfold greater affinity for this enzyme than does retin*ol* [30]. Thus, ethanol inhibits retin*al* generation within the testes and thereby may limit spermatogenesis [30].

The feminization seen in chronic alcoholic men is explainable less readily [6, 33]. As noted above, alcoholic men are phenotypically feminized as manifested by the presence of a female escutcheon and habitus, palmer erythema, spider angiomata and gynecomastia [28]. In addition, they are biochemically feminized as manifested by the presence of increased levels of sex steroid binding globulin, estrogen-sensitive neurophysin, prolactin growth hormone, thyroid stimulating the estrone... all of which are common to women and presumed to be estrogenic responses [32]. Several mechanisms have been proposed to explain these phenomena. First, because alcoholics with advanced liver disease are more likely to be feminized than are those without liver disease, the presence of liver disease per se has been advanced as the immediate cause of the feminization. Experimental support exists for such a supposition. First, alcoholic males with Laennec's cirrhosis have been shown to convert androstenedione to estrone and then to estradiol at a rate greater than do men without such liver disease [10]. Secondly, the presence of portal-systemic shunts has been shown to be associated with increased plasma estrone levels [27]. Presumably such shunts allow biliary excreted estrogens to be reabsorbed and then bypass the liver and reexcretion and thus stimulate estrogen-responsive peripheral tissues. Quantitatively, even more important, however, is the reabsorption of weak androgens such as androstenedione which are excreted into the bowel and are also reabsorbed, only to escape hepatic reexcretion as a result of the same portal-systemic shunts. Such reabsorbed androgens are converted to estrogens (principally estrone) at estrogen-responsive tissues without ever having circulated in the plasma as an estrogen. Finally, it has been shown recently that hypogonadism per se in the male reduces the hepatic content of an estrogen-binding protein that may protect the cell from excess estrogenic stimulation [8, 15]. As a result of primary alcohol-induced hypogonadism there is a secondary reduction in the amount of this estrogen-binding protein which consequently allows the cell to experience relatively more estrogen for any given estrogen level. This effect may be the most important mechanism responsible for feminization seen in men with advanced alcoholic liver disease.

The Alcoholic Female

In contrast to the male, the chronic alcoholic female is not superfeminized but instead shows severe gonadal failure commonly manifested by oligoamenorrhea, loss of secondary sexual characteristics such as breast and pelvic fat accumulation, and in addition, infertility [35]. Histologic studies of the ovaries obtained at autopsy from chronic alcoholic women, who have died of cirrhosis while still in their reproductive years (20–40 years of age), have shown a paucity of developing follicles and few or no corpora lutea, thus documenting reproductive failure [12]. Moreover, these findings have been reproduced in an animal model [9].

Endocrine failure of the ovary of alcoholic women is manifested by reduced plasma levels of estradiol, loss of secondary sex characteristics and ovulatory failure. The biochemical mechanisms for such endocrine failure are probably the same as those occurring within the testes of the male, as the pathways for steroidogenesis are the same in the gonads of the two sexes and an alcohol dehydrogenase has been reported to be present within the ovary.

In addition to demonstrating evidence of primary gonadal failure, chronic alcoholics, whether male or female, also show evidence of a central hypothalamic-pituitary defect in gonadotropin secretion [36]. Thus, despite severe gonadal injury, FSH levels although increased are well below the levels expected for the degree of reproductive failure present [28]. Further, despite the marked reduction in sex steroid levels, LH concentrations range from normal to only moderately increased. Moreover, many alcoholic men have been shown to have inadequate responses to both clomiphene and luteinizing hormone-releasing factor stimulation [28, 36].

Viral-Induced (Postnecrotic) Cirrhosis

In contrast to the considerable information evaluating hypothalamic-pituitary-gonadal function in individuals with alcohol-induced Laennec's cirrhosis, little data is available concerning hypothalamic-pituitary function in individuals with postnecrotic or viral-induced liver disease. In a recent evaluation of the hypothalamic-pituitary-gonadal axis of adult male hemophiliacs with advanced liver disease of presumed viral etiology, it was shown that testosterone levels were not reduced when compared to those of age-matched controls [38]. In addition, sperm concentrations were not reduced and ejaculates collected from such men were no different than those obtained from normal age-matched controls. Thus, in contrast to alcoholics, and despite comparable biochemical and histologically advanced liver disease, there is little evidence of prmary gonadal failure in men with advanced viral-induced liver disease.

LH concentrations, however, are increased in such men being roughly 2.5 times those of normal age-matched controls [38]. As expected, FSH levels are not increased. Moreover, when their gonadotropin responses to LRF are examined, such men are found to have normal gonadotropin responses. In addition, in response to clomiphene stimulation, hemophiliacs with advanced viral-induced liver disease have a normal twofold increase in the plasma levels of both gonadotropins [38]. Similarly, unlike the situation in alcoholics with liver disease, growth hormone levels are normal under basal as well as TRH-stimulated conditions. Moreover, basal and TRH-stimulated TSH levels are normal. Finally, in contrast to the findings in alcohol-induced cirrhosis, prolactin levels in hemophiliac individuals with presumed viral-induced liver disease are either reduced or normal under basal conditions and demonstrate a reduced response to TRH stimulation [38].

Based upon these findings, it has been suggested that despite quite similar degrees of histological and biochemical liver disease, individuals such as hemophiliacs with presumed viral-induced liver disease differ from individuals with an alcohol-induced liver disease in that those with viral-induced liver disease present evidence for a state of compensated gonadal dysfunction manifested by normal testosterone levels and elevated LH levels, and failure to demonstrate any evidence for hypothalamic-pituitary dysfunction [38]. To our knowledge, no data is available which critically evaluates hypothalamic-pituitary-gonadal function in women with viral-induced liver disease.

Hemochromatosis

In hemochromatosis, gonadal failure characterized by a high prevalence of reduced libido, impotence, reduction in sexual body hair and a reduction of testicular size has been reported to be present in as many as 50% of the male patients so evaluated. When evaluated critically, the pathophysiologic mechanisms responsible for these alterations in gonadal function have been less uniform than those seen in individuals with either alcohol or viral-induced liver disease [4, 22, 25, 42]. In about one-half of the reported studies, LH levels have been reduced suggesting secondary hypogonadism. In many others, however, LH levels have been increased suggesting primary gonadal failure. These variable results are not unexpected if one recalls that at autopsy, iron deposition is common in both the hypothalamus and pituitary and is occasionally seen also in the testes of men dying with hemochromatosis. Feminization is less common in individuals with hemochromatosis than in those with alcohol-induced liver disease presumably because portal-system shunting is less severe and/or the hypogonadism appears so much later in life that feminization does not become manifested except in unusual circumstances.

Probably because hemochromatosis is unusual in women, except in those well beyond the menopause, no data is available evaluating hypothalamic-pituitary-gonadal function in women with hemochromatosis.

Wilson's Disease

Unlike hemochromatosis, Wilson's disease is manifested in females as well as males and frequently is well advanced early in life. Little if any published data exists concerning hypothalamic-pituitary-gonadal function in individuals with Wilson's disease. We have had the opportunity to evaluate such function in only a few such patients and then only in men. The results of these few studies would suggest that hypogonadism is unusual in patients with Wilson's disease and when it occurs, it is usually associated with well-advanced, eventually lethal disease [40].

Therefore, our limited experience would suggest that hypothalamic-pituitary-gonadal dysfunction is quite unusual in such patients. Such a finding is not particularly surprising, however, when one recalls that copper is a known stimulant for all the pituitary hormones except prolactin [4]. This fact is even more meaningful when one recalls that of all the pituitary hormones only prolactin has an antigonadal function [17, 39]. Thus, increased central nervous system levels of copper commonly present in individuals with Wilson's disease would be expected to have elevated gonadotropin levels and reduced prolactin levels, factors that would be expected to maintain rather than inhibit normal gonadal function.

Biliary Cirrhosis (Both Primary and Secondary)

As with Wilson's disease increased hepatic copper levels are common in individuals with cirrhosis due to biliary tract disease whether it be primary or secondary [23, 27]. This is true presumably because the biliary system is the principal route for copper excretion in the body. If central nervous system copper levels are also increased in such diseases, as they are in Wilson's disease, then the same copper-induced hypersecretion of pituitary hormones would be expected regardless of any other influence upon the hypothalamic-pituitary-gonadal axis. Should other factors such as an interrupted enterohepatic circulation with portalsystemic shunts also exist, they would be expected to suppress gonadotropin secretion. Therefore, the net effect would be a normalization of plasma gonadotropin levels. Again in limited studies in such patients, this is exactly what we have observed [41]. Gonadal function is maintained and the entire hypothalamic-pituitary-gonadal axis appears to be intact and functioning normally.

Lupoid (Autoimmune) Chronic Active Hepatitis

Presumed autoimmune chronic active hepatitis is a disease of unknown etiology that is seen more commonly in women than men [43]. Clinically, it is associated with hirsutism, acne and maintenance of body muscle mass, all signs which suggest androgen excess. In addition, oligo- and/or amenorrhea is a common finding in such women. To our knowledge, no studies have been reported which have critically assessed the hypothalamic-pituitary-gonadal axis in such patients.

Many patients with this presumed autoimmune disease are found to also have other 'autoimmune' endocrine disease such as Hashimoto's thyroiditis, idiopathic hypoparathyroidism, and autoimmune adrenal and gonadal failure. The presumption that the failure of these endocrine tissues is due to an autoimmune mechanism is based principally upon the findings of (1) autoantibodies to the tissues of these endocrine organs; (2) a close association of these diseases with the HLA B8 and DW3 antigens, and (3) their occasional remission in response to immune suppression with drugs such as prednisone and azathiaprine. Despite such insights into the disease, little knowledge of hypothalamic-pituitary-gonadal function or dysfunction exists in such patients. Specifically, are the presumed excess androgen levels present in such individuals measurable? if so, what are they? and, finally, are they of adrenal or gonadal origin or both? The answers to these intriguing questions are not yet available.

As must be obvious from the above discussion, much remains yet to be learned about the effects of specific liver diseases and their etiologies upon the functioning of the hypothalamic-pituitary-gonadal axis. In the last decade progress in this area has been considerable; however, many questions still remain unanswered.

References

1 Baker, H.W.; Burger, H.G.; De Kretser, D.M.; Dulmanis, A.; Hudson, B.; O'Connor, S.; et al.: A study of the endocrine manifestations of hepatic cirrhosis. Q. Jl Med. 45: 145–178 (1976).
2 Barr, R.W.; Son, S.C.: Endocrine abnormalities accompanying hepatic cirrhosis and hepatoma. J. clin. Endocr. Metab. 17: 1017–1029 (1957).
3 Bennett, H.S.; Boggenstoss, A.H.; Butt, H.R.: The testes, breast and prostate of men who die of cirrhosis of the liver. Am. J. clin. Path. 20: 814–819 (1970).
4 Bezuoda, W.R.; Bothwell, T.H.; Vanderwalt, L.A.; Kronheim, S.; Pimstone, B.L.: An investigation into gonadal dysfunction in patients with idiopathic haemochromatosis. Clin. Endocrinol. 6: 377–385 (1977).
5 Brooks, R.V.: Androgens Clinics in endocrinology and metabolism. 4: 503–520 (1975).
6 Chopra, I.J.; Tulchinsky, D.; Greenway, F.L.: Estrogen-androgen imbalance in hepatic cirrhosis. Ann. intern Med. 79: 198–203.
7 Cobb, C.F.; Ennis, M.F.; Van Thiel, D.H.; Gavaler, J.S.; Lester, R.: Acetaldehyde and ethanol are direct testicular toxins. Surg. Forum 29: 641–644 (1978).
8 Eagon, P.K.; Fisher, S.E.; Imhoff, A.F.; Porter, L.E.; Stewart, R.R.; Van Thiel, D.H.; Lester, R.: Estrogen binding proteins of male rat liver: Influences of hormonal changes. Archs Biochem. Biophys. 201: 486–499 (1980).

9   Galvëo-Teles, A.; Burke, C.W.; Anderson, D.C.; Marshall, J.C.: Biologically active androgens and estradiol in men with chronic liver disease. Lancet *1973:* 173–177.
10 Gordon, G.G.; Olivo, J.; Rafii, F.; Southren, A.L.: Conversion of androgens to estrogens in cirrhosis of the liver. J. clin. Endocr. Metab. *40:* 1018–1026 (1976).
11 Gordon, G.G.; Altman, K.; Southren, A.L.; Rubin, E.; Lieber, C.S.: Effect of alcohol on sex hormone metabolism in normal men. New Engl. J. Med. *295:* 793–797 (1976).
12 Jung, Y.; Rossfield, A.B.: Prolactin cells in the hypophysis of cirrhotic patients. Archs Path. *94:* 265–669 (1972).
13 Kent, J.R.; Scaramuzzi, R.J.; Lauwers, W.; Parlow, A.F.; Hill, M.; Penardi, R.: Plasma testosterone, estradiol and gonadotropins in hepatic insufficiency. Gastroenterology *64:* 111–115 (1973).
14 La Bella, F.; Dulor, R.; Vivian, S.: Releasing or inhibiting activity of metal ions present in hypothalamic extracts. Biochem. biophys. Res. Commun *52:* 786–794 (1973).
15 Lester, R.; Eagon, P.K.; Van Thiel, D.H.: Feminization of the alcoholic. The estrogen/testosterone ratio (E/T). Gastroenterology *76:* 415–417 (1979).
16 Lloyd, C.W.; Williams, R.H.: Endocrine changes associated with Laennec's cirrhosis of the liver. Am. J. Med. *4:* 315–330 (1948).
17 Martin, J.B.; Reichlin, S.; Brown, G.M.: Regulation of prolactin secretion and its disorders in clinical neuroendocrinology, pp. 129–145 (Davis, Philadelphia 1977).
18 Morrione, T.: The effect of estrogen on the testes in hepatic insufficiency. Archs Path. *37:* 39–47 (1944).
19 Olivo, J.; Gordon, G.G.; Rafii, F.; Southren, A.L.: Estrogen metabolism in hyperthyroidism and in cirrhosis of the liver. Steroids *26:* 47–56 (1975).
20 Rather, L.J.: Hepatic cirrhosis and testicular atrophy. Archs intern. Med. *80:* 397–405 (1947).
21 Siiter, P.; Febres, F.: Ovarian hormone synthesis, circulation and mechanisms of action; in DeGroat, Cahill, Odell, Montini, Potts, Nelson, Steineburger, Winegrad, Endocrinology, vol. 3, p. 1406 (Grune & Stratton, New York 1979).
22 Simon, M.; Franchimont, P.; Murie, N.; Ferrand, B.; Van Lauwenberge, H.; Bourel, M.: Study of somatotropic and gonadotropic pituitary function in idiopathic haemochromatosis (31 cases). Eur. J. clin. Invest. *2:* 384–389 (1972).
23 Smallwood, R.A.; Williams, H.A.; Rosenoer, V.M.: Liver copper levels in liver disease. Studies using neutron activation analysis. Lancet *ii:* 1310–1313 (1968).
24 Southren, A.L.; Gordon, G.G.; Olivo, J.; Rafii, F.; Rosenthal, W.S.: Androgen metabolism in cirrhosis of the liver. Metabolism *22:* 695–702 (1973).
25 Stocks, A.E.; Powell, L.W.: Pituitary function in idiopathic haemochromatosis and cirrhosis of the liver. Lancet *i:* 298–300 (1972).
26 Tepperman, J.: Metabolic and endocrine physiology; 3rd ed. (Yearbook, Chicago 1973).
27 Van Berge, H.E.; Negonwen, G.P.; Tangedahl, T.N.; Hofmann, A.F.; Northfield, T.C.; La Russo, N.F.; McCall, J.T.: Biliary secretion of copper in healthy men: Quantitation by an intestinal perfusion technique. Gastroenterology *72:* 1228–1231 (1977).
28 Van Thiel, D.H.; Lester, R.; Sherins, R.J.: Hypogonadism in alcoholic liver disease. Evidence for a double defect. Gastroenterology *67:* 1188–1199 (1974).

29   Van Thiel, D.H.; Lester, R.: Sex and alcohol. New Engl. J. Med. *291:* 251–253 (1974).
30   Van Thiel, D.H.; Gavaler, J.S.; Lester, R.: Ethanol inhibition of Vitamin A metabolism in the testes: Possible mechanism for sterility in alcoholics. Science *186:* 941–942 (1974).
31   Van Thiel, D.H.; Gavaler, J.S.; Lester, R.; Goodman, M.D.: Alcohol induced testicular atrophy. An experimental model for hypogonadism occurring in chronic alcoholic men. Gastroenterology *69:* 326–332 (1975).
32   Van Thiel, D.H.; Gavaler, J.S.; Lester, R.; Loriaux, D.L.: Plasma estrone, prolactin, neurophysin and sex steroid binding globulin in chronic alcoholic men. Metabolism *23:* 1015–1019 (1975).
33   Van Thiel, D.H.; Lester, R.: Alcoholism. Its effect on hypothalamic-pituitary-gonadal function. Gastroenterology *71:* 318–327 (1976).
34   Van Thiel, D.H.; Lester, R.: Sex and alcohol. A second peek. New Engl. J. Med. *295:* 835–836 (1976).
35   Van Thiel, D.H.; Gavaler, J.S.; Lester, T.; Sherins, R.J.: Alcohol-induced ovarian failure in the rat. J. clin. Invest. *61:* 624–632 (1978).
36   Van Thiel, D.H.; Lester, R.; Vaitukaitis, J.: Evidence for a defect in pituitary secretion of luteinizing hormone in chronic alcoholic men. J. clin. Endocr. Metab. *47:* 499–507 (1978).
37   Van Thiel, D.H.; Gavaler, J.S.; Cobb, C.F.; Sherins, R.J.; Lester, R.: Alcoholic-induced testicular atrophy in the adult male rat. Gastroenterology *105:* 888–895 (1979).
38   Van Thiel, D.H.; Gavaler, J.S.; Slone, F.L.; Cobb, C.F.; Smith, W.I., Jr.; Bron, K.M.; Lester, R.: Is feminization in alcoholic men due in part to portal hypertension. A rat model. Gastroenterology *78:* 81–91 (1980).
39   Van Thiel, D.H.; Gavaler, J.S.; Spero, J.A.; Egler, K.M.; Wight, C.; Sanghvi, A.; Hasiba, U.; Lewis, J.H.: Patterns of hypothalamic-pituitary-gonadal dysfunction in men with liver disease due to differing etiologies (in preparation).
40   Van Thiel, D.H.: Unpublished observation.
41   Van Thiel, D.H.: Unpublished observation.
42   Walsh, C.H.; Wright, A.D.; Williams, J.W.; Holder, G.: A study of pituitary function in patients with idiopathic hemochromatosis, J. clin. Endocr. Metab. *43:* 866–872 (1976).
43   Wright, R.; Millward-Sadler, G.H.: Chronic active hepatitis of unknown etiology; in Wright, Alberti, Karran, Millward-Sadler, Liver and biliary disease, pp. 670–674 (Saunders, Philadelphia 1979).

Dr. David H. Van Thiel, 1000 J Scaife Hall, Department of Medicine,
University of Pittsburgh School of Medicine, Pittsburgh, PA 15261 (USA)

Neuroendocrinological Implications of Alcoholism

Arun K. Rawat

Alcohol Research Center, C. S. 10002, University of Toledo, Toledo, Ohio, USA

Introduction

It is becoming increasing clear that the central nervous system (CNS) plays a major role in regulating endocrine functions. The CNS not only controls the output of hormones, but it is itself a target tissue for these hormones. The peripheral hormones and even the peptide hormones of the anterior pituitary have actions on the brain which serve to regulate both the behavior and the subsequent hormone secretion. This is especially true for the steroid hormones, such as cortisol from the adrenal cortex and the androgens, estrogens, and progestins from the gonads, which rapidly pass through the blood-brain barrier and are selectively accumulated by specific cell groups in the brain. Since the most pronounced biochemical and pharmacological effects of ethanol are observed on the CNS, it is not surprising that renewed interest is developing in studying the ethanol-neuroendocrine interactions. It is through the biochemical and physiological studies in experimental models that the exact mechanisms of ethanol-mediated changes can be elucidated and the human pathology of alcoholism understood.

This study will examine the effects of short- and long-term ethanol administration on (1) the hormones and regulatory peptides of the anterior pituitary (adenohypophysis); (2) the hormones of the posterior pituitary (neurohypophysis) on putative neurohormones; (3) neurohormones in fetal alcohol syndrome, and (4) attempts will be made to specify the site of ethanol action and the metabolic consequences of such actions on neuroendocrine function.

It has often been said that the master gland in the brain is the pituitary which is located at the base of the brain. Hormones, such as

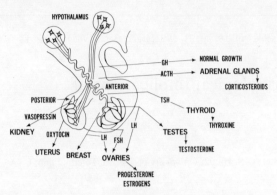

Fig. 1. Hormonal Inter-Relationship of Pituitary. The neuroendocrine relationship of anterior and posterior pituitary gland hormones to various hormones produced by other body organs.

adrenocorticotropic hormone (ACTH), thyroid-stimulating hormone (TSH), luteinizing hormone (LH) and follicle-stimulating hormone (FSH) from the anterior pituitary (adenohypophysis), control the secretion of hormones from the adrenal cortex, thyroid gland and gonads. Two other adenohypophysial hormones, namely prolactin and growth hormone (GH) exert their direct effect on the somatic tissues such as the mammary gland and the bones. The hormones of the posterior pituitary (neurohypophysis) have a variety of direct effects on tissues such as the uterus and the kidney. The posterior pituitary has also been shown to contain granules which hold oxytocin and vasopressin. In recent years it has been established that the brain indeed synthesizes and secrets chemicals which are carried through the blood stream to the anterior pituitary, where they cause the secretion of the tropic hormones. In 1970 the first of these releasing hormones was identified and synthesized, the releasing hormone was called thyrotropin-releasing hormone (TRH). This brain hormone causes the anterior pituitary to secret TSH. The effects of ethanol on these individual hormones will be examined in the following section (fig. 1).

Effects of Ethanol on Hypothalamic-Pituitary-Thyroid Axis

The number of studies concerning the acute or chronic effects of ethanol on thyroid function is limited. However, it has been shown that the ingestion of moderate amounts of ethanol (1.5 g/kg) had no effect on TSH in healthy male humans. TSH levels, after ethanol ingestion, were found to be similar to the control levels. Furthermore, the response of the pituitary to TRH injection was similar in the control and ethanol-consuming groups. In acutely ethanol-treated rats it has been observed that a significantly greater amount of ^{131}I is taken up following the injection of ^{131}I-triiodothyronine [4] by several tissues in rats. The thyroid gland of ethanol-treated rats showed a decreased level of ^{131}I. Chronic ethanol consumption in rats resulted in an increased ^{131}I uptake by the thyroid gland [39]. It is not clear from these studies whether the plasma levels of thyroxin were altered or not. Other studies [1] have determined the plasma levels of protein-bound iodine and triidothyronine in 103 acutely intoxicated male and female alcoholics and did not find any abnormality in thyroid function.

Although the exact mechanism of action of TRH is not known at present, it has been shown that intravenously or intraperitonially administered TRH antagonizes ethanol's narcosis and ethanol-induced hypothermia in the rodents [5, 6, 13], without affecting the rate of ethanol metabolism. It is interesting in this respect that in some clinical studies it has been reported that TRH lessens the depression associated with ethanol withdrawal syndrome [8]. However, the antidepressant action of TRH lasts only the day of its administration in human subjects experiencing ethanol withdrawal syndrome [8]. It has been suggested by several studies that TRH interacts with the neurotransmitters in the CNS. There is a strong possibility that TRH-induced effects on the CNS maybe mediated through the cholinergic systems [6, 10, 25].

Studies of the influence of thyroid hormone on the rate of ethanol metabolism have recorded contradictory observations. In the older literature enhancing effects of triiodothyronine on the rate of alcohol disappearance from the blood were reported [45]. However, other authors failed to find any changes in the oxidation rate of ethanol after triidothyronine administration [54]. In biochemical studies [54], attempts were made to investigate the possible influence of thyroxin on the metabolism of ethanol by liver and to understand the possible mechanisms of action of this hormone on ethanol metabolism. In these studies, it was observed that the rate of ethanol oxidation in liver slices from thyroxin-treated, methylthiouracil-fed and thyroidectomized rats were not significantly different from those of the livers of normal animals. Although thyroxin treatment did not significantly alter the oxidation-reduction state of the liver, thiouracil treatment and thyroidectomy

Table I. Influence of thyroxine administration, methyl–thiouracil feeding, and thyroidectomy on the rate of ethanol oxidation, and on lactate and pyruvate concentrations in the medium [from 54]

Treatment	Rate of ethanol oxidation μmol ethanol/g wet liver/h	Lactate/pyruvate
Control (injected with 0.01 N NaOH)	41.4 ± 5.7 (12)	87.0
Thyroxine	38.8 ± 4.4 (15)	88.7
Control (fed normal diet)	38.6 ± 3.8 (11)	89
Thiouracil-treated	42.1 ± 3.0 (6)	121
Thyroidectomized	38.7 ± 4.1 (5)	123

resulted in an increase in the hepatic cytoplasmic NADH/NAD ratio. However, under all these conditions the rates of ethanol oxidation were not significantly affected as shown in table I.

Effects of Ethanol on Hypothalamic-Pituitary-Adrenal Complex

Studies with mice and rats [27] have suggested that chronic ethanol consumption attenuated the hypothalamic-pituitary-adrenal response to ethanol. It has been shown [37] that serum cortisol is elevated in alcoholics following ethanol consumption. A possible alteration of the diurnal rhythm of hypothalamic-pituitary-adrenal complex has been observed in alcoholics [27]. In this study 3 alcoholics were studied who exhibited attenuated variation in the corticosteriod levels with two daily peaks rather than one normal peak. Other studies [38] have also shown higher morning levels of cortisol in alcoholics as compared to control subjects. Thus the available experimental data indicate that ethanol activates the hypothalamic-pituitary-adrenal system in various species of animals.

A number of stresses are known to activate the hypothalamic-pituitary-adrenocortical system which results in the release of glucocorticoids. In recent years direct measurements of the elevation of circulating glucocorticoids after ethanol administration have been used as an index of adrenocortical activation. Secretion of corticosterone following ethanol (2 or 4 g/kg) is inhibited in hypophysectomized rats or in pentobarbital/morphine-treated rats. Morphine is an inhibitor of ACTH release from the pituitary gland. Support for the notion that ethanol exerts its effect directly on the CNS comes from the injection of dexamethasone, a synthetic glucocorticoid, which inhibits the normal release of corticosterone after ethanol injection in mice. Furthermore, ethanol results in a significant reduction in pituitary ACTH content.

Effects of Ethanol on Hypothalamic-Pituitary-Gonadal Complex

Acute studies [59] with male rats have shown that a single injection of ethanol (ca. 2–2.5 g/kg) lowers the plasma LH. Depending on the dose, ethanol can have opposite effects of the levels of this hormone. Ethanol injections of less that 1 g/kg were stimulatory to LH and testtosterone (measured at 3 and 4 h, respectively), whereas injections of more than 1 g/kg were inhibitory [9].

In studies with humans [66], no change in plasma testosterone or LH levels was observed during the first few hours following ethanol ingestion (1.5 g/kg). However, approximately 15 h after cessation of drinking, there was a definite decrease in plasma testosterone and concurrent increase in LH.

In chronic studies with male rats, it has been observed that testicular, prostatic, and seminal vesicle atrophy occurs in addition to lowered plasma testosterone levels [63, 64]. Ethanol was given in a liquid ethanol diet in this study, with isocaloric sucrose controls for 20–60 days. The results showed that plasma testosterone was significantly decreased by 29–55% after ethanol administration for a period of up to 24 days. The decrease of plasma testosterone levels following ethanol intake was partially explained by the increased metabolism, which was attributed to the increased activity of hepatic testosterone-A-ring reductase, the rate-limiting enzyme in the metabolism of testosterone. Studies on alcoholics have often shown hypogonadism and hyperestrogenization in chronic alcoholic males. Lowered levels of testosterone in several male alcoholics [12] have been reported. On the basis of a prompt rise in testosterone following administration of human chorionic gonadotrophin to male alcoholics, it has also been proposed [58] that hypogonadism is probably secondary to hypothalamic-pituitary suppression. Subsequent studies have confirmed the lowered testosterone levels and normal responsiveness of alcoholics to human chorionic gonadotrophin [62]. However, it is not yet fully clear whether the impact of ethanol on the reproductive system is the initial site(s) of action of ethanol; whether it is a direct action on the testis or an inhibition of gonadotrophin release, which in turn leads to depressed testosterone levels. It is possible that ethanol might be altering another hormone, such as prolactin, which in conjunction with a relatively constant level of LH might alter testosterone levels. It is likely that the hormonal changes observed in alcoholics are in part the result of direct effects of ethanol on the CNS as well as peripheral target organs.

Impotence, testicular atrophy and gynecomastia have long been associated with ethanol-related liver disease [12]. Recent studies have suggested that some of these effects of chronic ethanol consumption occur in male alcohol addicts who show no signs of liver pathology or impaired hepatic function. Disturbances of endocrine function are believed to underlie these clinical problems, but little attention has been directed at elucidating the sites and mechanisms of ethanol action on the hypothalamic-pituitary-gonadal axis. In one experimental study [18] ethanol was found to suppress gonadotropin-stimulated testosterone biosynthesis in sexually mature male rats in vivo and in intact Leydig cells in vitro. A similar ethanol dose-response inhibition curve was also observed in this study, when either whole or homogenized cells were stimulated by dibutyryl cyclic-AMP. In the presence of pharmacologically relevant ethanol concentrations, added NAD^+ restored testosterone production to stimulated control levels. These results would suggest a direct inhibitory effect of ethanol on testicular testosterone synthesis. It would be logical to suggest from these studies that the site of inhibition is primarily intracellular and the mechanism is probably through a decrease in the $NAD^+/NADH$ ratio caused by ethanol oxidation.

Effects of Ethanol on Growth Hormone and Somatostatin

GH is released from the anterior lobe of the pituitary (adenohypophysis). In fact, this hormone has been observed to support growth in hypophysectomized animals. It has been shown that GH increases the output of glucose by the liver and the uptake of glucose by tissues. Plasma insulin activity is also increased, and the diabetes which results from GH treatment may be due to the increased output of insulin and subsequent exhaustion of the beta cells of the pancreas. The effects of GH upon fat metabolism appear to be mediated through the adrenocortical and thyroid hormones, an early effect of the administration of the hormone being a mobilization of carcass fat to the liver followed by an increased fat breakdown and increased production of ketone bodies.

Table II. Effect of insulin administration, alloxan diabetes and anti-insulin serum treatment on the ethanol oxidation rate by liver slices [from 46]

Treatment	Additions	Rate of ethanol oxidation μmol ethanol/g wet weight/h
Control (injected with saline)	nil	43.8 ± 1.3 (6)
	glucose (11 mM)	44.9 ± 1.8 (6)
	fructose (11 mM)	51.0 ± 2.5 (6)
Insulin-treated (5 U insulin/rat in 48 h)	nil	59.6 ± 2.8 (6)
	glucose (11 mM)	63.8 ± 3.2 (6)
	fructose (11 mM)	44.5 ± 2.1 (6)
Alloxan diabetic (60 mg/kg body weight i.v.)	nil	35.4 ± 1.7 (6)
	glucose (11 mM)	36.8 ± 2.3 (6)
	fructose (11 mM)	37.5 ± 2.6 (6)
AIS treatment (1.0 ml guinea pig AIS/rat)	nil	35.1 ± 1.5 (6)
	glucose (11 mM)	36.4 ± 1.9 (6)
	frutose (11 mM)	37.0 ± 2.7 (6)

In healthy human volunteers who fasted overnight, a significant and sustained elevation of GH after oral administration of ethanol (1.2 g/kg in a 20% solution) has been observed [2]. A concurrent, but slightly delayed decrease in free fatty acids and an increase in plasma 11-hydroxycorticosteriods was also observed in these subjects. Insulin-induced hypoglycemia is a potent stimulus for the release of GH in humans. In normal human volunteers, ethanol (diluted with water) was administered 30 min prior to an insulin injection [27]. In the presence of ethanol, the GH response to insulin-induced hypoglycemia was significantly reduced in terms of its peak height and duration of response. Since there was no difference in the hypoglycemic effect following insulin injection in the presence or absence of ethanol, it was concluded in this study that ethanol exerted a direct attenuating effect on the brain centers controlling GH secretion.

Most of the studies on the effect of ethanol on GH levels have been done on human subjects. According to one study [21], there have been over 100 cases of ethanol-induced hypoglycemia. This has been suggested to be a direct inhibitory effect of ethanol on gluconeogenesis without either an increase in the insulin level or a major hepatic malfunction. The homeostatic regulation of glucose is very complex, and controlled by a network of many hormones, such as insulin, glucagon, GH, and adrenal hormones.

Table III. Effect of glucagon on the rate of ethanol oxidation in rat liver slices [from 47]

Treatment	Additions	Rate of ethanol oxidation µmol/g/h	Lactate/pyruvate
Control	nil	nil	46.6
	ethanol	45.6 ± 4.2 (4)	73.0
Glucagon, (30 g followed by 5 µg)	nil	nil	66.6
	ethanol	65.0 ± 6.5 (4)	80.0
Glucagon, (200 g followed by 30 µg)	nil	nil	73.0
	ethanol	4010 ± 4.8 (4)	98.7
Control	glucagon	nil	53.0
	ethanol + glucagon	44.8 ± 4.3 (4)	78.0

Although it is known that both insulin and glucagon affect the rates of ethanol metabolism, the interplay of GH in these effects is not quite clear. Especially when there is a fine balance among GH, insulin and glucagon. Some of the interactions of ethanol, insulin and glucagon in the cortex with glucose homeostasis will be discussed in the following section. Insulin administration to normally fed male rats was observed [46] to result in a significant increase in the rate of hepatic ethanol metabolism. Conversely, administration of anti-insulin serum (AIS) or alloxan led to a decrease in the rate of ethanol metabolism by hepatic tissue (table II). Administration of low doses of glucagon (30 µg followed by 5 µg) has also been observed [35] to result in a significant increase in hepatic ethanol metabolism as shown in table III. This effect of glucagon was not observed either at high doses (200 µg followed by 30 µg), or upon in vitro addition of glucagon to liver tissue [47]. It was concluded from these studies that, at low doses, glucagon may be exerting its effect through stimulation of insulin secretion. The details of carbohydrate metabolism, hormonal effects, their effects on ethanol metabolism are beyond the scope of this article, the reader is referred to an detailed review in this area [45].

Somatostatin, a brain peptide has been found in the extra-hypothalamic region of several animals. Somatostatin is regarded as peptidergic neurotransmitter and presumably acts directly on the CNS. Studies with somatostatin indicate that its administration results in reduced motor activity, hypothermia, enhanced pentobarbital-induced sleep time and enhanced pentobarbital-induced mortality [11, 42, 56]. Although an excitatory effect of somatostatin on the CNS has been observed [43], most of the evidence cited above suggests an inhibitory influence of somatostatin on the CNS. Although several investigations are available regarding the interactions of somatostatin and the depressants of CNS, much information is not available about the interactions of ethanol and somatostatin. However, it is becoming increasingly plausible that somatostatin might be involved in some of the symptoms of alcohol withdrawal, such as tremors, body rigidity and seizures [36].

Effects of Ethanol on Vasopressin and Oxytocin

It has been suggested [16] that acute ethanol administration leads to a depression in the activity of posterior pituitary gland and it was further suggested that this may be the cause of ethanol-induced diuresis. It was further observed [17] that the ethanol-induced diuresis in man could be antagonized by the extract of the posterior pituitary. Other studies [45] suggested that ethanol acted on the supraoptic nucleus to block the release of vasopressin. This suggestion is further strengthened by observations where it was observed that intracarotid infusion of ethanol to dogs produced a prompt diuresis at a time when the periphereal blood ethanol levels were not detectable. It was concluded from the studies that the diuretic effect of ethanol is due to the central inhibition of antidiuretic hormone. The retention of sodium, potassium, and chloride during ethanol-induced diuresis, has been demonstrated.

The effect of ethanol administration on the hemorrhage-induced increase of aldosterone secretion in pentobarbital-anesthetized dogs has been examined [20]. Under normal experimental conditions it was observed that a biphasic increase in the aldosterone response to terminal hemorrhage occurs. Alterations in the secrection rate of aldosterone have been observed when the animals were infused with varying amounts of ethanol in saline 1 h prior to initiation of hemorrhaging. Blood ethanol levels of 100 mg/dl stimulated aldosterone secretion.

In studies with normal male humans, it has been observed [27] that aldosterone decreases during the intoxication phase and increase during the withdrawal phase. On the other hand, plasma renin levels increase during both the intoxication and withdrawal phases. The changes in urinary Na^+/K^+ ratios agree with the plasma aldosterone levels during ethanol intoxication and withdrawal.

Ethanol inhibits the release of the posterior pituitary hormones oxytocin and vasopressin (antidiuretic hormone) and this accounts for well-known diuretic effect of alcohol. Recently oxytocin has been shown to be released from the maternal pituitary in the pulsatile fashion throughout labor, and inhibition of this release by ethanol has been demonstrated [49].

Oxytocin release as determined by uterine motility was inhibited by ethanol [23]. The effect of ethanol on milk ejection in the rat was studied [33]. It was demonstrated that although the latency between the onset of nursing and milk ejection was increased following an injection of 5 g ethanol/kg. The average time between milk ejections and the milk yields were about the same in the ethanol-treated and control animals. It might be considered that the eventual oxytocin release is an adaptation of the central effect to the inhibiting property of ethanol. This inhibition of oxytocin release was tested clinically to prevent premature labor.

Treatment of threatened premature labor with ethanol is based on experimental evidence for the inhibition of oxytocin secretion from the neurohypophysis by administration of ethanol [22]. Since the beginning of the century it has been assumed that oxytocin plays a role in the initiation and maintenance of labor. Studies based upon bioassays gave some support for this assumption, but it was not until the development of more sensitive and specific radioimmunoassays that evidence was provided for the presence of oxytocin in maternal and fetal blood during labor in concentrations adequate to stimulate the uterus to contract [10, 14].

Endorphins: Opiate-Like Peptides of Brain or Pituitary

The generic name endorphins derived from endogenous morphine was proposed by *Eric Simon*. The discovery of opiate receptors in the brains of mammals led to the demonstration of several of these oligopeptides from the pig hypothalamus-neurophysis with opiate activity [41]. Biochemical and histochemical studies have suggested that these opiate-like peptides are neurotransmitters in the CNS. It has been shown that several of these peptides can inhibit the excitability of a variety of neurones in the CNS. Cerebrospinal fluid injection of endorphins affects several behavioral and physiological measures [7, 34, 35] as well as responses to noxious agents [3]. β-Endorphins induce a morbid catatonic state which lasts for hours.

Since the endorphin field is new and developing, not much attention has been given so far to their interactions with ethanol, or the response of the body to ethanol intoxication, alcohol dependence or alcohol withdrawal as it relates to endorphins. The need for intensive scientific investigation in this area is obvious to understand the mechanism of ethanol intoxication, alcohol dependence and alcohol withdrawal. However, studies by *Blum* are beginning to suggest that the lower the encaphalins the higher the desire to drink.

Possible Interactions between Peptidergic and Putative Neurohormones

The possible mechanisms of action for peptidergic hormones are not defined. There are several available hypotheses; however, they require more rigorous scientific documentation. Unfortunately, even less is known with regard to the biochemical events that lead to observed behavioral changes following administration of these hormones. For example it has been suggested [19] that the CNS actions of somatostatin may be mediated in part through cholinergic mechanisms. Cerebral infusions of somatostatin (25–50 μg) induced a behavioral response referred to as 'barrel rotation', whereas intraperitoneal injection of atropine prior to or during rotation blocked these somatostatin-induced changes [19]. Pretreatment of rats with haloperidol, reserpine, or apomorphine (drugs that alter dopamine-mediated activity) had no effect on somatostatin-induced barrel rotations. Therefore a single somatostatin-induced behavioral change (barrel rotation) requires a functional cholinergic system, whereas a similar TRH-induced change in locomotor activity apparently operated through a dopaminergic mechanism.

Since the postsynaptic actions of several putative neurotransmitters involve cyclic AMP as a second messenger, changes in cyclic AMP levels in the neostriatum, cerebral cortex, and hippocampus following intracerebroventricular infusion of somatostatin have been examined [61]. Cyclic AMP levels increased in the cerebral cortex, neostriatum, and hippocampus at 5 and/or 15 min after the intracerebroventricular infusion of somatostatin (10 μg). Sotalol (β-adrenergic blocker) pretreatment eliminated or lessened somatostatin-induced increases in cyclic AMP in the hippocampus and cerebral cortex but not in the neostriatum [61]. However, the behavioral effects of intracerebroventricular-injected somatostatin appear not to involve cyclic AMP as the second messenger, since the typical locomotor excitation induced by somatostatin was unaffected by pretreatment with sotalol, whereas cyclic AMP levels were reduced [19].

Effects of Ethanol on Putative Neurohormones

Putative neurohormones such as acetylcholine, γ-aminobutyric acid (GABA), and biogenic amines (e.g., norepinephrine, dopamine, and serotonin = 5-hydroxytryptamine) play an important role in the CNS by acting as transmitters of neuronal functions. In view of the importance of neurotransmitter substances, a large number of studies have been conducted to investigate the effects of ethanol or acetaldehyde on their metabolism, uptake, turnover, and storage. In this section the relationship between ethanol and neurotransmitters is discussed.

Steady-State Levels of Neurohormones

The effects of ethanol and its metabolite, acetaldehyde, have been studied in the past in several species. However, the results obtained by various investigators are quite often not directly comparable. However, it is clear from these studies that the observed variation in the effects of ethanol can be largely attributed to differences in dosage, route of administration, duration of the study, and areas of the brain studied.

Effect on Norepinephrine and Serotonin

Intravenous administration of ethanol (2 g/kg) to rabbits was reported [62] to cause a significant decrease in norepinephrine and serotonin content in the brain stem up to 8 h later. In rat whole brain, however, an increase in serotonin content was observed 1 h after ethanol administration [49].

Since the steady-state concentration of neurotransmitters in the brain is very susceptible to postmortem changes such as hyperthermia and anoxia, a study was designed to rule out such possibilities [55]. The effects of intraperitoneally administered ethanol (3 g/kg) on whole-brain norepinephrine and serotonin levels were studied. Control animals were given saline solution, and after 15 min of ethanol administration the animals were sacrificed by total immersion in liquid nitrogen. Studies on such frozen brain preparations showed that ethanol did not result in a significant change in the steady-state concentration of serotonin and norepinephrine. It appears from these studies that intraperitoneally or orally administered ethanol has little effect on steady-state concentrations of serotonin and norepinephrine.

In one study [49] in which ethanol was given for over a period of 2 months as a 10% solution ad libitum, it was observed that serotonin increased significantly in the cerebellum, mesencephalon, and rhombencephalon. It was also observed in this study that, whereas dopamine levels did not change significantly in any of the above-mentioned areas of the brain, the norepinephrine content decreased. In another study on mice [48] the effect of chronic ethanol administration on the steady-state concentrations of norepinephrine and serotonin was investigated. It was found that, whereas chronic ethanol administration did not significantly alter the concentration of serotonin in the brain, the norepinephrine content showed a significant decrease. The levels of norepinephrine returned to control values after 4 days of ethanol withdrawal. A consistent observation in the above-mentioned studies was a decrease in the content of norepinephrine in the brain as a result of chronic ethanol administration.

Effect on GABA and Enzymes of Its Metabolism

Both acute (3 g/kg) and chronic prolonged (4 weeks, 6% w/v) administration of ethanol resulted in a significant increase in the concentration of GABA in the adult mouse [64]. The observed increase in the content of GABA in the brain has been suggested to be due to an increase in the content of glutamate, since these two amino acids are metabolically interrelated. However, in some studies a decrease in rat brain GABA content was observed [49]. However, the results of this study cannot be attributed to the action of ethanol, since the GABA content was determined in the brains of rats decapitated at room temperature. As mentioned earlier, the steady-state concentration of neurotransmitters is very susceptible to postmortem hypoxia and hyperthermia, and therefore in determining in vivo levels of GABA the use of rapid freezing of the tissue is essential.

Effect on Dopamine and Enzymes of Its Metabolism

The role of dopamine as a neurotransmitter in the CNS has been discussed in detail elsewhere [52]. Although inhibition of monoamine oxidase (MAO) by alcohol has been studied [53], not much work has been done in this regard on the metabolism of dopamine as a substrate for MAO. Homovanillic acid (HVA) is the major metabolic product of dopamine. Ethanol administration has been observed to decrease the excretion of HVA without causing a significant effect on excretion of the alcohol derivative of this metabolite, namely, 3-methoxy-4-hydroxyphenyl ethanol (MOPET). However, there is not much information on the effect of ethanol on the levels of dopamine in vivo in the brain.

Unfortunately, the work done regarding the effect of ethanol on dopamine turnover in the brain is not quite consistent. Chronic administration of ethanol for 5 days has been observed to cause a decrease in dopamine turnover in the brain. No significant effect of ethanol on dopamine turnover was reported in one study [52]. It seems that contrasting observations have been made in this regard as a result of the use of different doses of ethanol and differences in the duration of the experiment. A dose-dependent effect of ethanol on brain levels of dopa was observed. Since no effect of ethanol was observed on the levels of serotonin in this particular study, it was suggested that ethanol may have a specific action on the catecholamine-metabolizing systems [52].

Effects on Histamine and Enzymes of Its Metabolism

Recent studies with pharmacological agents [44, 60], subcellular fractions of the brain [8, 28], and lesion experiments [57] have suggested a neurotransmitter role for histamine in the CNS. Available developmental data and histochemical studies suggest histamine in the brain may be located in the synaptic vesicles of the nerve terminals, and in the non-neuronal mast cells [24]. Although, the mast cells are rare in the CNS, in certain rodent species they have also been found to exist in the CNS [15].

Histamine in the brain has been shown to influence diuresis, thermoregulation, and EEG activity; functions which are also affected by ethanol [65].

It was observed in one study [51] that acute administration of ethanol or acetaldehyde after 20 min resulted in a significant increase in brain histamine levels. This increase in brain histamine levels seems to be a direct result of brain histamine release due to ethanol or acetaldehyde metabolism in the body. Chronic consumption of ethanol also resulted in an increase in histamine levels in the brain [51].

Effect on Acetylcholine and Enzymes of Its Metabolism

The role of acetylcholine in central synaptic transmission has been well documented, and most studies are highly suggestive of a role as a central neurohumoral transmitter. It is also well known that the levels of acetylcholine in the brain vary with the functional activity of the organ [49] and are affected by a variety of drugs.

Our studies on mouse brain showed that acute administration of either ethanol (3 g/kg i.p.) or acetaldehyde (40 mg/kg i.v.) resulted in a decrease in the cerebral concentration of acetylcholine 15 min after administration. Chronic (4 weeks, 6% v/v) administration of alcohol also resulted in a decrease in whole-brain acetylcholine content in the mouse [55]. The effects of ethanol or acetaldehyde on cerebral acetylcholine content are consistent with the effect of these substances on cerebral acetyl-CoA and CoA content. The levels of both these compounds are decreased in the brain by ethanol or acetaldehyde. Acetylcholine content in the brain was also observed to decrease after ethanol administration [49]. However, it is difficult, to ascertain whether the observed changes seen in this study can be attributed to ethanol or the ether anesthesia and the technique of sacrificing the animals at room temperature by cervical dislocation. In contrast to the in vivo effects of ethanol, incubation of brain cell preparations or potassium-stimulated brain slices does not affect acetylcholine concentration or its release [49].

Ethanol intoxication also interferes with the incorporation of choline into acetylcholine. Although the steady-state concentration of cerebral choline is not affected by acute or chronic administration of ethanol, a decreased incorporation of (^{3}H)-choline into cerebral acetylcholine from acute ethanol-treated (3 g/kg) mice was observed as reported [48]. Although acute ethanol administration did not affect the activity of the enzymes involved in acetylcholine synthesis and degradation, chronic (4 weeks, 6% v/v) ethanol feeding of mice resulted in a decrease in cerebral acetylcholine transferase activity [48].

Effects on Rate of Turnover of Neurotransmitters

Studies on the rate of turnover of serotonin have produced conflicting results regarding the effects of ethanol on this presumptive neurotransmitter. It was observed in one study [26] that ethanol administration (3.3 g/kg i.p.) resulted in a small decrease in serotonin turnover. Using mice as experimental animals [29], it was found that acute ethanol administration (4 g/kg i.p.) did not result in a significant change in serotonin turnover. However, chronic administration of ethanol in a liquid diet (2 g/kg/day) in mice was found to result in a significant increase in serotonin turnover at 8 and 14 days. An increase in the activity of tryptophan hydroxylase activity was also observed in this study. However, a decrease was observed in serotonin turnover after acute and chronic ethanol administration in intoxicated rat brains. From the observations cited above it seems that the effects of ethanol on serotonin turnover are totally inconclusive, and no sound conclusions can be drawn from the existing data.

The effects of ethanol on the turnover of norepinephrine are less controversial and there is greater agreement that acute or chronic ethanol administration results in an increased turnover of norepinephrine. However, it is not clear whether this effect is a direct or an indirect one.

Uptake of Putative Neurohormones

A large number of studies have been conducted to investigate the effects of ethanol or acetaldehyde on the uptake of various neurotransmitters by brain tissue in vitro. When a very high ethanol concentration (0.22 M) was used, it was observed that the norepinephrine uptake by brain slices decreased [49]. The uptake of several presumptive neurotransmitters by synapsis when studied [49], showed that only the uptake of glutamate was affected and that the uptake of norepinephrine, serotonin, and GABA was unaffected. However, it is difficult to offer an explanation regarding the mechanism of the action of ethanol.

Urinary Excretion of Biogenic Amines

Several studies have demonstrated that the ingestion of ethanol by humans results in alterations in the urinary excretion of certain biogenic amines and their metabolites, although in such studies the concentrations of biogenic amines were determined in the urine, representing an overall picture of biogenic amine metabolism in the brain. It was observed that ingestion of 0.71 g ethanol/kg resulted in a significant increase in the urinary excretion of norepinephrine, metanephrine, and dopamine [49]. A significant decrease in the excretion of 5-hydroxyproline acetic acid was observed. Other studies have shown that ethanol produces increase in urinary excretion of catecholamines and their metabolites in animals [49].

It seems safe to assume that there is increased excretion of certain biogenic amines. This increased excretion may result from adrenal activation. The question which remains to be answered is: What is the relation of this increased excretion of biogenic amines to intoxication, alcohol addiction, and alcohol withdrawal?

Neurohormones in Fetal Alcohol Syndrome

Recent studies from this and other laboratories have shown that prolonged ethanol exposure, either during gestation or lactation, exerts a number of injurious and toxic effects on the developing fetus and the newborn [30, 31]. In humans the condition was described by *Lemoine* et al. [32]. Since then these studies have been confirmed and expanded by several independent investigators [40, 50, 52]. Human studies have shown that children born to chronic alcoholic mothers show developmental delay, growth deficiency and behavioral and neurological abnormalities. This condition has been termed as 'fetal alcohol syndrome' (FAS). Although, the neurochemical mechanisms of the FAS are not known, some studies are becoming available at the neurotransmitter level in the FAS. It is, however, important to state that although, changes in the neurotransmitter levels are observed in FAS, it is not claimed that these changes are responsible for the observed neurological abnormalities. As of now, more scientific investigation is needed to establish a direct cause-effect relationship.

Neuroendocrine Function in FAS

Since attention to the biochemical abnormalities in alcoholic fetopathy or FAS has been directed only recently, much work is not available on the endocrinological alteration in FAS.

Recently information about the effects of ethanol administration on sex hormone metabolism in normal men has been acquired. These include an augmented conversion of androgenic precursors to estrogens, a higher plasma concentration and production rate of estradiol and an increased plasma concentration of estrone. The alteration of metabolism of androgens has been shown to be coupled with direct effects of ethanol on hypothalamic-pituitary function, with changes in plasma LH, as described earlier in this article. However, no such data on corresponding changes in women have been published. This may be an important factor in FAS. Women with chronic liver disease who drink heavily might endogenously produce hormones which alter fetal cardiovascular and genital development.

Hypothalamic-pituitary function in four children born to an alcoholic woman has been studied to determine if hormonal abnormalities account for FAS aberrant growth patterns. In one study [62], severe postnatal growth deficiency was investigated in five cases of FAS. It was observed following insulin-induced hypoglycemia and arginine infusion that there was a slight hyperresponse to GH and normal somatomedin activity. Elevation of GH levels in the presence of growth failure has been previously reported in cases of severe malnutrition and in rare cases of pituitary dwarfism in which peripheral tissue insensitivity to GH has been postulated [62].

In studies with rats, pregnant animals have been kept on a liquid diet containing ethanol (6.8% v/v) from the 13th day of gestation, serum GH levels have been determined in the offspring [61]. In neonates exposed to ethanol throughout development, serum GH, secretion was initially found to be increased and then declined to subnormal levels compared to controls. The levels returned to normal after weaning. Withdrawal from ethanol at birth produced similar alterations in serum GH concentrations to those seen in neonatal rats exposed to ethanol continuously. Postnatal exposure to ethanol alone also caused a decrease in GH concentration in rat pups. These data indicate that exposure to ethanol during gestation causes an alteration in the regulation of GH secretion in the developing pups. A single dose of ethanol given to pups postnatally also caused a marked decline in serum GH levels demonstrating that in the developing pup ethanol can interfere acutely with GH release. If these findings can be replicated, one could speculate that exposure to ethanol in utero may have caused reduction of the number of cells in the peripheral tissue of FAS children.

Neurotransmitters in FAS

One of the frequently observed occurrences in alcoholic women is breech birth. It is tempting to suggest that, because of alterations in brain inhibitory and excitatory neurotransmitters, the neurological signs in the baby become abnormal. This may lead to a breech birth. In an attempt to understand the mechanism, at the molecular level, which might result in such abnormal neurological signs, a study was undertaken in our laboratory [4, 49]. It was observed that chronic ethanol feeding of pregnant rats resulted in alterations in the steady-state concentration of several neurotransmitters in the fetal brain. Chronic ethanol consumption by pregnant animals on a Metrecal diet resulted in a significant increase in the cerebral content of GABA in the fetal brain at various ages of gestation. The cerebral glutamate content was also found to be increased in this study. Although the cerebral serotonin content was not significantly altered, the norepinephrine concentration showed a significant increase in the ethanol-fed group.

Ethanol consumption by lactating mothers also resulted in significant changes in the steady-state concentration of several presumptive neurotransmitters in the neonatal brain. Neonates suckling on alcohol-fed mothers were examined from 0 to 12 days after birth. As in the case of fetuses, in neonates ethanol also led to an increase in the cerebral content of GABA, glutamic acid, and norepinephrine [4, 52]. However, in contrast to the results obtained with fetal brains, in neonatal brains the steady-state concentration of serotonin was increased until 10 days after birth, after which time it started to return to normal. Chronic ethanol consumption by pregnant or lactating animals resulted in a significant decrease in the steady-state concentration of acetylcholine in the brains of both fetuses and neonates. In addition to measurements of the steady-state concentration of neurotransmitters in fetal and neonatal brains, the effects of ethanol on some of the enzymes involved in the synthesis and breakdown of neurotransmitters were also examined. The activity of glutamate decarboxylase and GABA transferase in fetal as well as neonatal brains was significantly decreased in the alcohol-fed group as compared to the sucrose-fed group. Although, the steady-state levels of neurotransmitters are important, there is an urgent need for the investigation of the turnover rates of these neurotransmitters in the fetal and neonatal brains. However, such studies are not yet available.

Conclusion

In conclusion, both short- and long-term ethanol consumption have significant effects on hormones and hormone-releasing factors from the anterior pituitary and posterior pituitary. This in turn influences both the behavior and subsequent hormone secretion from other organs of the body. Ethanol not only influences neurohormones, it also directly affects the neurotransmitters and therefore the delicate interhormonal balance is influenced. It has become increasingly clear in recent years that several of the actions of peptidergic neurohormones may in fact be mediated through the neurotransmitter substances. Ethanol, by affecting both these parameters independently, is capable of changing their metabolic influences on the body. Last but not least, prolonged maternal alcoholism during pregnancy leads to hormonal and neurotransmitter abnormalities which may be responsible for the aberrant growth and neurological anomalies patterns observed in FAS.

Summary

Short- and long-term consumption of alcohol has significant effects on the endocrine systems. These effects are especially important since ethanol affects neuroendocrine systems, which in turn alter the balance of brain secretions from pituitary and the target organs. The implications of ethanol-influenced alterations in peptidergic hormones, putative neurohormones, and CNS functions have been examined in this article.

References

1 Augustine, J.R.: Lab studies in acute alcoholics. Can. med. Ass. J. 96: 1367–1370 (1967).
2 Bellet, S.; Yoshimine, N.; Decastre, O.A.P.; Roman, L.; Parmer, S.; Sandberg, H.: Effect of ethanol administration on plasma growth hormone response to insulin-induced hypoglycemia. Metabolism 20: 762–769 (1971).
3 Belluzzi, J.D.; Grant, N.; Garsky, V.; Sarantakin, D.; Wise, C.C.; Stein, L.: Analgesia introduced in vivo by central administration of enkephalins in rat. Nature, Lond. 260: 625–626 (1976).
4 Bleecker, M.; Ford, D.H.; Rhines, R.K.: A comparison of ^{131}I-triiodothyronine accumulation and degradation in ethanol R and control rats. Life Sci. 8: 267–275 (1969).
5 Breese, G.; Cott, J.; Cooper, B.; Prange, A., Jr.; Lipton, M.; Plotnikoff, N.: Effects of TRH on the actions of pentobarbital and other centrally acting drugs. J. Pharmac. exp. Ther. 193: 11–22 (1975).
6 Breese, G.; Cott, J.; Cooper, B.; Prange, A., Jr.; Lipton, M.; Plotnikoff, N.: Interaction of thyrotropin-releasing hormone with central acting drugs in Prange, the thyroid axis, drugs and behavior, pp. 115–127 (Raven Press, New York 1974).
7 Bloom, F.; Segal, D.; Ling, N.; Guillemin, R.: Endorphins: profound behavioral effects in rats suggest new etiological factors in mental illness. Science 154: 630–632 (1976).

8   Carlini, E.; Green, J.P.: The subcellular distribution of histamine, slow-reacting substance and 5-hydroxytryptamine in the brain of the rat. Br. J. Pharmacol. *20:* 264–277 (1963).
9   Cicero, J.J.; Badger, T.M.: Effects of alcohol on the hypothalamic pituitary-gonadal axis in the male rat. J. Pharmac. exp. Ther. *201:* 427–433 (1977).
10   Chard, T.; Body, N.R.H.; Forsling, M.L.; NcNeilly, A.S.; Landon, J.: The development of a radioimmunoassay for oxytocin. J. Endocr. *48:* 223–230 (1970).
11   Cohn, M.L.; Cohn, M.: 'Barrel' rotation induced by somatostatin in the non-lesion rat. Brain Res. *96:* 138–141 (1975).
12   Coppage, W.L.; Cooner, A.E.: Testosterone in human plasma, androgen metabolism in the cirrhosis of liver. New Engl. J. Med. *273:* 902–907 (1965).
13   Cott, J.; Breese, G.; Cooper, B.; Barlow, T.; Prange, A., Jr.: Effects of TRH on the actions of pentobarbital and other centrally acting drugs. J. Pharmac. exp. Ther. *196:* 594–604 (1976).
14   Dawood, M.Y.; Raghavan, K.S.; Posiask, C.; Fuchs, F.: Oxytocin in human pregnancy and parturition. Obstet. Gynec., N.Y. *51:* 138–143 (1978).
15   Dropp, J.J.: Mast cells in the central nervous system of several rodents. Anat. Rec. *174:* 227–238 (1972).
16   Edkins, N.; Murray, M.M.: Sugar tolerance and alcohol. J. Physiol., Lond. *71:* 403–411 (1931).
17   Eggleton, M.G.: The diuretic action of alcohol in man. J. Physiol., Lond. *101:* 172–191 (1942).
18   Ellingsboe, J.; Varanelli, C.C.: Ethanol inhibits testerone biosynthesis by direct action on leydig cells. Res. Commun. chem. Pathol. Pharmacol. *24:* 87–102 (1979).
19   Eskay, R.L.: Interaction of thyrotropin-releasing hormones and somatostatin with ethanol and other sedative/hypnotics. In Majchrowicz, Noble, Biochemistry and Pharmacology of ethanol (Plenum Press, New York 1979).
20   Fabre, L.F.; Farmer, R.W.; Pillizzari, E.D.; Farrell, G.L.: Aldosterone secretion in pentobarbital-anesthetized ethanol-infused dogs. Quart. J. stud. alc. *33:* 476–481 (1972).
21   Freinkel, N.; Singer, D.L.; Arkey, R.A.; Bleicher, S.J.; Anderson, J.B.; Silbert, C.K.: Alcohol hypoglycemia. I. J. clin. Invest. *42:* 1112–1133 (1963).
22   Fuchs, A.R.; Fuchs, F.: The possible mechanisms of labor inhibition by ethanol; in Josimovich, Uterine contraction (Wiley, New York 1973).
23   Gibbens, G.L.; Chard, T.: Observations on maternal oxytocin release during human labor and the effect of intravenous alcohol administration. Am. J. Obstet. Gynec. *125:* 243–246 (1976).
24   Green, J.P.: Histamine; in Lajtha, Handbook of neurochemistry, vol. 4, pp. 221–250 (Plenum Press, New York 1970).
25   Horita, A.; Carion, M.I.; Chesnut, R.M.: Influence of TRH on drug induced narcosis and hypothermia in rabbits. Psychopharmacol. Bull. *49:* 57–62 (1976).
26   Hunt, W.A.; Majchrowicz, E.: Turnover rates and steady-state levels of brain serotonin in alcohol-dependent rats. Brain Res. *72:* 181–184 (1974).
27   Kakihana, R.; Butte, J.C.: Ethanol releasing and endocrine function; in Majchrowicz, Noble, Biochemistry and Pharmacology of ethanol (Plenum Press, New York 1979).
28   Kuhar, M.J.; Taylor, K.M.; Snyder, S.H.: The subcellular localization of histamine and histamine methyltransferase in rat brain. J. Neurochem. *18:* 1515–1517 (1971).

29 Kuriyama, K.; Rauscher, G.E.; Sze, P.E.: Effect of acute and chronic administration of ethanol on the 5-hydroxytryptamine turnover and tryptophan hydroxylase activity of the mouse brain. Brain Res. *26:* 450–454 (1971).

30 Jones, K.L.; Smith, D.W.; Streissguth, A.P.; Myrianthopoulos, N.C.: Outcome in offspring of chronic alcoholic women. Lancet *i:* 1076–1078 (1974).

31 Jones, K.L.; Smith, D.W.; Hanson, J.W.: The fetal alcohol syndrome: clinical delineation. Ann. N.Y. Acad. Sci. *273:* 130–137 (1976).

32 Lemoine, P.; Harousseau, H.; Borteyru, J.P.; Menuet, J.C.: Les enfants de parents alcooliques. Anomalies observées. A propos de 127 cas. Ouest méd. *25:* 477–482 (1968).

33 Lincoln, D.W.: Milk ejection during alcohol anaesthesia in the rat. Nature, Lond. *243:* 227–229 (1973).

34 Linkola, J.; Fyrhquist, F.; Nieminen, M.M.; Weber, T.H.; Tontti, K.: Endorphins: profound behavioral effects in rats suggest new etiological factors in mental illness. Eur. J. clin. Invest. *6:* 191–198 (1976).

35 Loh, H.H.; Tseng, L.F.; Wei, E.; Li, C.H.: Beta-endorphin a potent analgesic agent. Proc. natn. Acad. Sci. USA *73:* 2895–2898 (1976).

36 Majchrowicz, E.: Induction of physical dependence upon ethanol and the associated behavioral changes in rats. Psychopharmacologia *43:* 245–254 (1975).

37 Mendelson, J.H.; Osata, M.; Mello, N.K.: Adrenal function and alcoholism. Serum cortisol. Psychosom. Med. *33:* 147–157 (1971).

38 Merry, J.; Marks, V.: Effects of alcohol ingestion and growth hormone. Lancet *ii:* 990–991 (1972).

39 Murdock, H.R.: Thyroid effects of alcoholism. Q. Jl Alcohol Stud. *28:* 419–423 (1967).

40 Ouellette, E.M.; Rosett, H.L.: A pilot prospective study of the fetal alcohol syndrome at the Boston City Hospital. II. The infants. Ann. N.Y. Acad. Sci. *273:* 123–129 (1976).

41 Pert, C.B.; Snyder, S.H.: Opiate receptor: demonstration in nervous tissue. Science *179:* 1011–1014 (1973).

42 Prange, A.J., Jr.; Breese, G.; Jahnke, G.; Cooper, B.; Cott, J.; Wilson, I.; Lipton, M.; Plotnikoff, N.: Parameters of alteration of phenobarbital response by hypothalamic polypeptides. Neuropsychobiology *1:* 121–131 (1975).

43 Plotnikoff, J.; Kastin, A.; Schally, A.: Growth hormone release inhibiting hormone: neuropharmacological studies. Pharmacol. Biochem. Behav. *2:* 693–696 (1974).

44 Pollard, H.; Bischoff, S.; Schwartz, J.C.: Increased synthesis and release of (^{3}H) histamine in rat brain by reserpine. Eur. J. Pharmacol. *24:* 399–401 (1973).

45 Rawat, A.K.: Influence of thyroid hormone on the rate of alcohol metabolism. In Influence of hormones and other factors on hepatic alcohol metabolism (Kandrup & Wunch, Copenhagen 1969).

46 Rawat, A.K.: Effect of hyper- and hypoinsulinism on the metabolism of ethanol in rat liver. Eur. J. Biochem. *9:* 93–100 (1969).

47 Rawat, A.K.: Effect of glucagon on the metabolism of ethanol in rat liver. Acta chem. scand. *24:* 1163–1167 (1970).

48 Rawat, A.K.: Brain levels and turnover rates of presumptive neurotransmitters as influenced by administration and withdrawal of ethanol in mice. J. Neurochem. *22:* 915–922 (1974).

49 Rawat, A.K.: Neurochemical consequences of ethanol on the nervous system. Int. Rev. Neurobiol. *19:* 123–172 (1976).
50 Rawat, A.K.: Effect of maternal ethanol consumption on foetal and neonatal rat hepatic protein synthesis. Biochem. J. *160:* 653–661 (1976).
51 Rawat, A.K.: Development of histaminergic pathways in brain as influenced by maternal alcoholism. Res. Commun. chem. Pathol. Pharmacol. *27:* 91–104 (1980).
52 Rawat, A.K.: Neurotoxic effects of maternal alcoholism on the developing fetus and newborn; in Manzo, Advances in neurotoxicology (Pergamon Press, Oxford 1980).
53 Rawat, A.K.: Pharmacological and toxicological considerations in fetal alcohol syndrome; in Messiha, Tyner, Alcoholism: a perspective (PJD Publications, New York 1980).
54 Rawat, A.K.; Lundquist, F.: Effects of thyroxine on glycerol and ethanol metabolism in rat liver slices. Eur. J. Biochem. *5:* 13–17 (1968).
55 Rawat, A.K.; Kuriyama, K.; Mose, J.: Metabolic consequences of ethanol oxidation in brains from mice chronically fed ethanol. J. Neurochem. *20:* 23–33 (1973).
56 Segal, D.S.; Mandell, A.J.: Differential behavioral effects of hypothalamic polypeptides. In Prange, The thyroid axis, drugs and behavior, pp. 120–133 (Raven Press, New York 1974).
57 Schwartz, J.D.; Julien, C.; Feger, J.; Garbard, M.: Histaminergic pathway in rat brain evidenced by hypothalamic lesions. Fed. Proc. *33:* 285 (1974).
58 Southren, A.L.; Gordon, G.G.; Olivo, J.; Rafii, F.; Rosenthal, W.S.: Androgen metabolism in the cirrhosis of liver. Metabolism *22:* 695–702 (1973).
59 Symon, A.M.; Markes, V.: Effects of alcohol on weight-gain and hypo-thalamic-pituitary-gonadotrophin access in maturing male rat. Biochem. Pharmac. *24:* 955–958 (1975).
60 Taylor, K.M.; Snyder, S.H.: Histamine in rat brain: sensitive assay of endogenous levels, formation in vivo and lowering by inhibitors of histidine decarboxylase. J. Pharmac. exp. Ther. *179:* 619–633 (1971).
61 Thadani, P.V.; Schanberg, S.M.: Effects of maternal ethanol ingestion on serum growth hormone in the developing rat. Neuropharmacology *18:* 821–826 (1979).
62 Tze, W.J.; Friessen, H.G.; MacLeod, P.M.: Growth hormone response in fetal alcohol syndrome. Archs Dis. Childh. *51:* 703–706 (1976).
63 Van Thiel, D.H.; Gavaler, J.S.; Lester, R.; Goddman, M.D.: Alcohol induced testicular atrophy: an experimental model for hypogonadism occurring in chronic alcoholic men. Gastorenterology *69:* 326–332 (1975).
64 Van Thiel, D.H.; Lester, R.; Sherins, R.J.: Hypogonadism in alcoholic liver disease; evidence for a double defect. Gastroenterology *67:* 1188–1199 (1974).
65 Wolf, P.; Monnier, M.: Electronencephalographic, behavioral and visceral effects of intraventricular infusion of histamine in the rabbit. Agents Actions *3:* 1976 (1973).
66 Ylikahri, R.; Huttunen, M.; Hardomen, M.; Seuderling, U.; Onikki, S.; Haronen, S.L.; Adlercreutz, H.: Low plasma testosterone values in man during hangover. J. Steroid Biochem. *5:* 655–658 (1974).

Prof. A.K. Rawat, Alcohol Research Center, C.S. 10002, University of Toledo, Toledo, OH 43699 (USA)

Effect of Ethanol on Spontaneous and Stimulated Growth Hormone Secretion[1]

Geoffrey P. Redmond

Departments of Pharmacology and Pediatrics, University of Vermont College of Medicine, Burlington, Vt., USA

Growth hormone (GH) is essential for statural growth in infants and children. It is an important anabolic hormone as is insulin; both peptides stimulate amino acid uptake into cells and induce net protein synthesis resulting in nitrogen retention. In contrast to insulin, GH elevates blood glucose; deficient individuals are vulnerable to hypoglycemia. It is possible that altered GH secretion might contribute to the metabolic derangements seen in chronic alcoholics. While the clinical picture is not that of acromegaly or GH deficiency, hormonal alterations combined with other effects of ethanol might mediate some of the abnormalities of chronic alcoholism. Growth failure is a major feature of the fetal alcohol syndrome. Accordingly, the actions of ethanol on GH secretion are of considerable interest.

Significance of Spontaneous Secretion

Until the last decade, endocrine physiology was dominated by the concept of negative feedback. The pituitary was thought to respond to changes in target hormone levels by a reciprocal change in secretion of the trophic hormone resulting in relatively constant levels of thyroxine, cortisol, testosterone, and, in the early part of the menstrual cycle, estrogen. That such an elaborate apparatus should exist simply to maintain

---

[1] Supported, in part, by grants 1R03MH33177-01 and 1R03AA03804-01 from the National Institute of Mental Health and National Institute of Alcohol Abuse and Alcoholism.

unchanging levels of these hormones, or indeed, that hormones should exist at all if their levels remained constant was not seriously questioned. With the introduction of radioimmunoassay (RIA) by *Berson and Yalow,* accurate hormone levels could be determined easily and quickly on minute volumes of blood. Levels could be measured frequently in the same individual permitting detailed characterization of secretion patterns. It soon became apparent that levels were not constant but varied in patterns characteristic for each hormone. An early approximation was the concept of a 'diurnal rhythm' for cortisol which was observed to be at high levels in the early morning hours but low or undetectable at night. This rhythm is actually the summation of a more complex pattern in which levels are determined by the frequency of discreet secretory episodes which are probably quantal in nature. For other hormones such as GH, the pattern is considerably less regular and more complex [12].

While the negative feedback model does describe certain aspects of hormone secretion, most secretion results from a preprogrammed secretory drive located in the brain and is not primarily determined by input from the periphery. This new concept is more satisfying teleologically. If the role of hormones is taken to be coordination of diffuse metabolic processes, then steady-state hormone levels cannot provide any dynamic regulation. However, changes in levels can signal distant tissues to alter their activity in response to a change in the external or internal environment. Brain activity is rhythmic (the sleep-wake cycle is the most evident example) and the hypothalamic-pituitary unit provides an effector organ by which the activities of distant organs can be coordinated. In this concept, the endocrine system has a functional role similar to that of the sympathetic nervous system.

GH Secretion in Man

Many drugs and other stimuli have been found to provoke GH release in man or in other primates. A detailed review of this subject is available elsewhere [9]. Exercise and physical or psychologic stress have been found to stimulate hGH release as have arginine and certain other amino acids, pituitary peptides, and a long list of drugs including epinephrine, *L*-dopa, apomorphine, bromocriptine, clonidine, and 5-hydroxytryptophan. Propranolol is often included in lists of drugs which stimulate

GH release but it probably blocks certain kinds of inhibition rather than actually stimulating release. In the rat, propranolol does not increase spontaneous rGH secretion [*Webb and Redmond,* unpubl. observations]. Although dopamine was once thought, to be a major stimulatory transmitter for GH release, it is now thought based on nonhuman primate studies, that norepinephrine occupies this role [2, 21]. Phentolamine infusion decreased the amplitude of secretory episodes but did not alter rhythmicity [21]. Inhibition by dopamine and/or somatostatin is probably involved in generating the secretory rhythm. Thus, antisomatostatin antibodies result in a lack of rhythmicity but integrated GH concentrations were not altered [20].

Several studies have examined spontaneous hGH secretion over 24 h. The general strategy is to place a catheter in the antecubital vein to permit subsequent sampling without pain or production of any other sensation in the patient. One approach is to draw out a sample every 20 min for 24 h [6]. An alternative is to remove blood continuously with a portable withdrawal pump. In this method, blood is usually collected in 30-min segments; individual determinations represent integrated rather than instantaneous hormone levels [13]. Although there are some differences in the results obtained by these two groups using slightly different methods, the patterns described are simular. As in the rat, secretion is episodic but secretory episodes are less frequent and less regular. The most characteristic feature is a major peak occurring at the onset of stage 3–4 sleep (slow wave sleep – SWS). In the study of *Finkelstein* et al. [6], there were no other secretory peaks achieving levels greater than 5 ng/ml in prepubertal children. Adolescents secreted a considerably greater amount of hGH and had several daytime and other nighttime peaks in addition to the early sleep related one. Young adults were intermediate with fewer peaks than adolsescents but more than prepubertal children. Older adults aged 47–52 did not all secrete measurable amounts of GH and in the two who did, one had a peak during sleep, the other, during waking hours. Because a total of only 5 older subjects were studied, firm conclusions are not possible. However, there is considerable other evidence that GH secretion declines in older individuals.

In contrast, *Plotnick* et al. [13] found daytime peaks in prepubertal subjects. A substantial peak after the onset of sleep was again noted. Visual inspection of their data suggests that peaks were higher but not more frequent in pubertal compared to prepubertal children. Adults in their twenties had fewer episodes but magnitude was in the higher range seen in adolescents. Their 24-hour secretory rate was lower than both prepubertal and pubertal children.

The differences in the results reported in these two studies may well be due to differences in study conditions. The subjects of *Plotnick* et al. [13] carried out normal activities on the ward whereas those of *Finkelstein* et al. [6] were confined to bed. Of necessity in such detailed studies, only a limited number of subjects can be included and it seems likely that the full range of variation in hGH secretion in normal subjects has not been fully revealed. The following conclusions are consistent with the results of both groups: (1) hGH secretion is episodic but regularity is not evident when individual results are examined. (2) A sleep-related peak occurs consistently in children, adolescents and young adults but not older adults. (3) hGH secretion lessens with increasing age after adolescence. (4) A limited number of samples over a period of less than 24 h do not give a meaningful picture of hGH secretory activity.

Readers interested in a more detailed consideration of hGH secretory patterns are referred to a recent review [12].

Although a number of studies have sought ethanol effects on hGH secretion, none used the 24-hour frequent sampling approach. Accordingly, their findings must be interpreted with great caution. Those studies known to the author, their findings and their limitations are reviewed in the following section.

Ethanol and GH

Human Studies

A least eight human studies have been published in which effects of ethanol on hGH secretion were sought. The results are in seeming conflict with *Arky and Freinkel* [1] and *Bellet* et al. [3] reporting apparent increases, *Othmer* et al. [10] and *Toro* et al. [25] reporting no quantitative change and the others reporting decreases [7, 8, 14, 23]. Many of these studies were performed at a time when knowledge of the physiology of anterior pituitary secretion was considerably less than it is today. The purpose of the following review is to explore the factors that account for disparate results in an effort to develop a unified concept of the effects of ethanol on GH secretion.

Arky et al. [1] were primarily interested in patients thought to have isolated deficiency of ACTH. To document that hGH secretion was normal in these individuals, hGH levels were measured by producing hypoglycemia. This was and remains the most reliable of the various stimulation tests for hGH release. Only 2 subjects were studied. Hypoglycemia was produced by insulin and by ethanol (150 ml of absolute alcohol diluted in saline over 8 h, i.v.) on separate occasions. GH rose to 30 and 40 ng/ml in the 2 patients after insulin but only 15 and 6 after ethanol. However, the conclusion that ethanol blunted the hGH response to hypoglycemia is not justified because blood glucose fell more rapidly and reached lower nadirs after insulin than after ethanol. Because the stimuli differed in intensity, it cannot be concluded that the ethanol was the factor blunting the response. Presumably, the increase in hGH observed with ethanol infusion is due to the resulting hypoglycemia and not to the ethanol itself. However, the data in this experiment alone do not permit proof of this assumption. The one normal studied was a 34-year-old woman who was fasted 3 days prior to ethanol infusion. Before ethanol she had a normal serum glucose but an hGH level of 22 ng/ml representing an appropriate elevation in response to fasting. Although her hGH levels fell after ethanol infusion, the lack of a control (e.g. saline infusion after a 3-day fast) makes it difficult to interpret the experiment. However, it is plausible that the fall in hGH was due to ethanol in light of the results of some of the other studies reviewed below.

Priem et al. [14] studied the effect of ethanol on the hGH response to hypoglycemia. Subjects served as their own controls and were given insulin or insulin plus ethanol (50 ml diluted in water) on separate occasions. All 7 subjects had lower peak hGH levels after ethanol than after insulin alone and the difference was significant ($p < 0.05$). Since there was a trend for lower blood glucose levels when ethanol was given, these results do indicate a blunting of the hGH response to hypoglycemia in the presence of ethanol. Two cautions need to be stated however. First, hGH levels on both occasions reached peaks well above the 5–7 ng/ml range used to define the upper limit achieved by hGH-deficient patients. It is possible, therefore, that the lower level after ethanol is no less effective than the higher one seen after insulin alone. Secondly, the change in hGH need not be a direct pharmacological action of ethanol exerted in the hypothalamus. As the authors themselves point out, free fatty acids (FFA) fell less after insulin in the presence of ethanol and the higher FFA may have blunted hGH secretion rather than ethanol itself. Certainly the implication of specific changes in hypothalamic monoamines as the mechanism of ethanol's actions is entirely speculative.

Infusion of the amino acid arginine is another stimulus commonly used to assess hGH secretion. Since hGH stimulates amino acid uptake into cells, hGH release in response to this might have some physiological relevance. However, spontaneous secretory episodes in normal human children and adults do not bear a clear relationship to meals [13]. Furthermore, the amount of arginine infused (usually 0.5 g/kg) greatly exceeds the amount contained in any normal meal. Hence, the relation of anginine-stimulated release to physiological release remains uncertain. *Tamburrano* et al. [23] gave 30 g anginine infused intravenously over 30 min to 3 normal and 4 acromegalic subjects. 1 week later an identical infusion of arginine was given following 75 g of ethanol which had been infused over 4 h. The hGH response was quite clearly blunted by prior ethanol in both normals and acromegalies. In normals after 30 min of arginine it was 8.4 ± 1.2 ng/ml in the control study and 2.8 ± 1.2 ng/ml with ethanol. Levels were generally lower with ethanol in the acromegalics. Since secretion from functioning pituitary tumors remains subject to control mechanisms to a limited degree, the ability of ethanol to blunt the hGH rise in acromegaly is not entirely unexpected. Of even greater interest in this study is the finding of a decline in hGH levels in acromegalic patients after ethanol during the 4 h before arginine was begun. This effect was noticeable 30 min after the start of ethanol and significant at every subsequent sample time suggesting inhibition of the small amount of secretion that was occurring. In normal subjects, hGH levels were very low in all periods prior to arginine so no light is shed on ethanol effects on episodic secretion – no secretory episode occurred during the study period.

Ganda et al. [7] examined the hGH response to propranolol-glucagon. Chronic alcoholics (defined as individuals consuming more than 1 qt whiskey/day or the equivalent) had a peak response of 8.9 ± 7.0 (SD) ng/ml, chronic alcoholics abstinent for 2–3 weeks had a peak of 18.5 ± 8.9 ng/ml; and a similar group of abstinent alcoholics given 300 mg of chlordiozepoxide over the 24 h preceding testing had a mean peak of 12.7 ± 7.4 ng/ml. The author states that the mean in normal men is 15.1 ± 6.5 ng/ml. These differences are rather small. For purposes of data analysis, the nonabstinent alcoholics were divided into responders and nonresponders. Peak values in nonresponders were significantly lower than those in other groups. Presumably, the first group did not differ from the others when not divided in this fashion although this is not explicitly stated. This manipulation of the data seems hard to justify and engenders scepticism about the conclusion that hGH secretion in nonabstinent alcoholics differs from the other groups. An additional problem is the failure to specify the extent and timing of alcohol use in the nonabstinent group in relation to administration of propranolol-glucagon.

Toro et al. [25] gave ethanol in the modest dose of 1 g/kg to normal men and women and reported that levels of hGH, cortisol, testosterone, LH and prolactin did not change. This abstract does not specify hGH levels; it is possible that levels were similar in both groups because no spontaneous secretory episodes occurred in either control or ethanol phases of the study since sampling covered only a 3-hour period. Changes in cortisol, testosterone and LH after ethanol have been well established. Failure to observe them in this study may have been due to the low dose of ethanol employed.

Three studies examined the effect of ethanol on spontaneous hGH secretion although none covered a 24-hour period. *Leppaluoto* et al. [8] studied 9 normal volunteers on two occasions, once with ethanol 1.5 g/kg diluted in 'beverage' of unspecified composition and once with beverage only. Ethanol was consumed orally between 6 and 9 p.m. Blood samples were taken at 6, 8, 10 and 12 p.m. and then at 7 and 9 a.m. Apparently, separate venipunctures rather than an indwelling catheter were used for sampling. Controls had a peak at 10 p.m (mean 3.1 ± 1.3 ng/ml) but the subjects who consumed oral ethanol did not (mean 0.6 ng/ml). Controls also had slightly higher levels at 8 p.m. but hGH levels at all other times were negligibly low in both groups. This study does, therefore, provide some evidence that ethanol inhibits spontaneous hGH secretion. Ethanol appeared to prevent the episode altogether rather than to alter its amplitude. Due to the infrequency of sampling, it cannot be ruled out that ethanol delayed the episode rather than preventing it completely.

An 'uncoupling' of SWS and hGH secretion in male alcoholics was reported by *Othmer* et al. [10]. Since the report is an abstract, few details are given. Alcoholics, apparently whether recently abstinent or given a fifth of bourbon, had hGH elevation after the onset of sleep but in contrast to normals, the EEG at the time of secretion was not the slow wave pattern. Since no numbers are given, little can be concluded from this study: the coordination of hGH secretion with sleep is not as regular in adults as in children and adolescents [6].

A single study has reported a rise rather than a fall in hGH levels following ethanol [3]. 11 healthy males were given oral ethanol in a dose of 1.5 g/kg as a 20% solution in water. On another occasion, several days apart, plain water was given. At 30 min, hGH levels were significantly higher after ethanol than in the control phase; the peak, which occurred at 90 min, was 5.94±5.73 ng/ml compared to control levels of 1.32± 1.40 ng/ml. FFA levels fell after ethanol but the fall followed the rise in hGH levels. Blood glucose and insulin were unchanged. There was a rise in 11-hydroxycorticoids, an endocrine change consistently found after ethanol. It is difficult to integrate these findings with those reported by others. The experiment appears well designed and carefully conducted. The study which is most comparable in design is that of *Leppaluoto* et al. [8]. They used the same dose of ethanol and studied normal young males. Ethanol prevented a spontaneous secretory episode observed under control conditions and did not elevate hGH at any time point sampled. However, the sample times in the two experiments cannot be directly compared. The subjects of *Leppaluoto* et al. consumed ethanol between 6 and 9 p.m. and samples were taken at 6, 8, 10 and 12 p.m. Thus, the precise time in relation to ethanol consumption is uncertain. Because of ethanol's zero order elimination kinetics, peak and integrated blood concentrations are much lower when absorbtion is gradual rather than rapid. Hence, the results of *Bellet* et al. [3] are not necessarily contradicted by those of *Leppaluoto* et al. [8]. Nonetheless, the preponderance of evidence favors the conclusion that ethanol inhibits, rather than stimulates, hGH secretion. The possibility of a brief rise in secretion preceding a phase when secretion is suppressed cannot be excluded in the present state of knowledge.

From the foregoing review, the following points emerge: (1) What evidence exists suggests that the major effect of ethanol on hGH release in man is inhibition. Only a single study found an increase and it was modest in magnitude and brief in duration. (2) All the reported studies suffer from important limitations, namely: (a) numbers of subjects are too small to support firm conclusions; (b) stimulation tests used in most studies tell little about secretion under physiological circumstance; (c) those studies examining spontaneous secretion did not sample long enough or frequently enough to adequately characterize changes in secretion.

Two related questions concerning ethanol effects on secretion of hGH (or any hormone) are pertinent and remain to be answered by a suitably designed human study: (1) Is the 24-hour integrated serum concentration of hGH altered by ethanol given acutely to normals and is it altered in chronic alcoholics? Do substantial changes occur at moderate blood levels or only at those obtained with very heavy drinking? (2) How does ethanol affect secretory characteristics? Are timing, frequency, amplitude and duration of secretory episodes altered?

Stimulation studies have two basic applications, clinical diagnostic studies and investigation of control mechanisms. For the latter, it is essential to use highly selective agents. Most available agents do not meet this criterion making interpretation of results uncertain. For example, oral L-dopa produces hGH release and it has generally been assumed that dopamine is a stimulatory transmitter for hGH release. More recently, evidence has appeared that suggests that norepinephrine, not dopamine, is the monoamine with a stimulatory role in hGH secretion [5].

Studies in the Rat

GH secretion in the rat, as in man, is episodic. In contrast to the pattern in man, however, secretion in the rat has been reported to be regular with major secretory episodes beginning at approximately 200-min intervals [24]. Because of the greater regularity of secretion in the rat, a full 24-hour sampling period is not necessary; however, several hours are still required to obtain meaningful data. Because there may be both individual variations in secretory pattern and also variations related to time of day, dual controls are desirable. Each rat is sampled for 3 h before ethanol or other treatment. In addition, other rats are given saline rather than ethanol at 3 h.

We have employed a chronically implanted carotid artery catheter to perform frequent sampling on rats in a nontraumatic fashion [15, 16]. Samples were taken every 15 min for a 6-hour period. All animals received infusions through the catheter of saline, red cells and heparin at 90, 180 and 270 min. Treated animals received ethanol in addition at 180 min. Hence, control and ethanol-treated animals did not differ in the number and time of intra-arterial injections. The methodology and its rationale are more fully described elsewhere [15, 16].

Results in some representative controls are shown in figure 1. In spite of variation between individual animals, there is a tendency for a

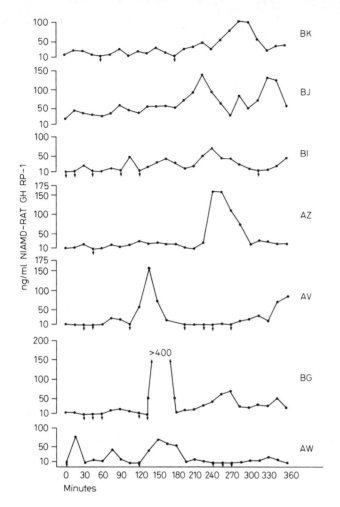

Fig. 1. Secretion of rGH in saline controls. rGH levels were determined every 15 min for 6 h starting at 9 a.m. Animals were heparinized with 200 U prior to sampling. At 90, 180 and 270 min each animal received an infusion consisting of its own previously removed red cells, saline equal in volume to plasma removed and heparin, 150 U. Initials identify individual rats [from ref. 15].

major secretory episode in each of the two 180-min periods. Not all animals show this, however. Peak r (rat) GH levels in ng/ml are much higher than hGH levels in man; the peptides are chemically distinct and rGH is inactive in man. The secretory pattern is similar to that reported by *Tannenbaum and Martin* [24] but somewhat less regular. When given

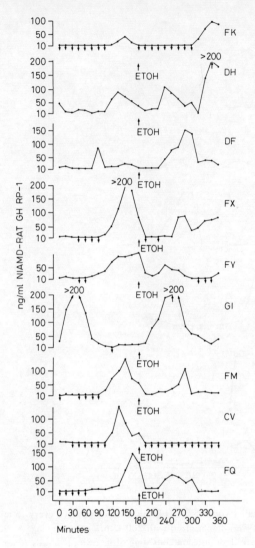

Fig. 2. Effect of ethanol 2 g/kg on rGH secretion. Conditions were identical to controls except for addition of ethanol to the 180-min infusion. There is no change attributable to ethanol. Initials identify individual rats [from ref. 15].

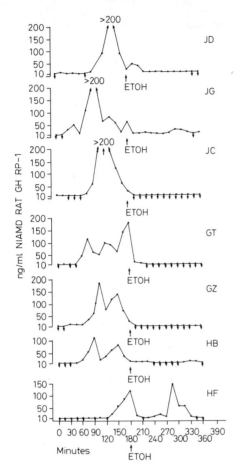

Fig. 3. Effect of ethanol 4 g/kg on rGH secretion. Conditions were identical to controls except for addition of ethanol to the 180-min infusion. Ethanol at this high dose significantly ($p < 0.001$) suppressed rGH secretion. Initials identify individual rats [from ref. 15].

in a dose of 0.5 or 2 g/kg, ethanol had no apparent effect on hGH secretion (fig. 2). However, doses of 3 and 4 g/kg totally abolished secretory activity in the 180 min following the dose (fig. 3). The duration of suppression of rGH secretion was not determined in these experiments. However, in studies of another drug which blocks rGH secretion, the phenothiazine perphenazine (5 ng/kg), suppression lasted more than 3, but less than 22 h [16]. There is no reason to believe that there is permanent damage to rGH secretory mechanisms after ethanol and presumably secretion resumes sometime after 3 h. Similarly, the effect of repeated or chronic dosing was not studied. Inhibition of rGH secretion after perphenazine is still seen in animals who have received the drug daily for the 30 preceding days [*Soncrant and Redmond*, unpubl. observations]. The author is aware of no precedent for specific tolerance to endocrine effects of a drug but the question cannot be definitely answered for ethanol on the basis of present knowledge. Since alcoholics adjust their dose by effect rather than by blood level, much of the tolerance to various actions of ethanol is overcome by increasing intake.

The dose-response relationship of rGH inhibition by ethanol merits some comment. First, it is of a threshold nature. No change occurred after 2 g/kg, but secretion stopped completely after 3 g/kg. Secretory episodes are either normal in amplitude and frequency or they cease completely. *Ellis* [3a] reported ethanol levels of 215 and 430 mg/dl in the rat after 2 and 4 g/kg. Presumably, then, the threshold blood level for rGH suppression is about 300 mg/dl. The animals in our study receiving doses of 2 g/kg or more were profoundly sedated, indeed anesthetized. Several of those given 4 g/kg died. Inhibition of rGH secretion must be considered an effect of high doses. At lower doses secretion remains intact even though many other aspects of nervous system function are altered. It seems likely, therefore, that rGH suppression is simply one aspect of the generalized depression of brain activity engendered by large amounts of ethanol rather than a specific or selective effect. Other general anesthetic agents such as pentobarbital similarly result in diminished secretion of several pituitary hormones.

GH Secretion in the Fetal Alcohol Syndrome (FAS)

Impaired body and brain growth are major manifestations of the FAS [11, 22]. For this reason, there has been some interest in hGH secretion in this disorder. Unfortunately, the human study reported by *Tze* et al. [26] has been widely misinterpreted. In this study, 5 female children with the FAS ranging in age from 0.4 to 7.6 were subjected to arginine and insulin stimulation tests on two separate days. No subject was deficient by the usual clinical criterion in that all had one or more values exceeding 7 ng/ml. The authors take the rather high levels they obtained in some individuals (for example 147.0 ng/ml after insulin in subject 3, a 2.3-year-old girl) as evidence of abnormally high hGH levels. Unfortunately, the study is uncontrolled and cannot support this inference. Elevation of hGH secretion is seen in Laron dwarfism, a syndrome in which resistance to the actions of hGH is manifested by failure to generate somatomedin in response to endogenous or exogenous hGH. Attractive as it is to regard the growth deficit in the FAS as due to resistance to hGH, this remain hypothetical at the present time.

There are other problems which diminish confidence in the findings and interpretations offered in this study. Subject 4 was born at 40 weeks gestation with a clearly normal birth weight of 3,040 g. Since intrauterine growth retardation is an essential feature of the syndrome, the diagnosis is in doubt. Similarly, 2,150 g (subject 5) is a normal birth weight for a 36-week infant. The details of the RIA for hGH were not stated; presumably, they were done as routine clinical specimens. Clinical laboratories frequently report spuriously high results of RIAs because of failure to truncate the standard curve at the shoulder. Most assays for hGH cannot accurately quantitate levels above about 40 ng/ml without special dilutions. This consideration is not mentioned in the article making it difficult to determine how much reliance can be placed on the actual values enumerated. The one conclusion that does seem firm is that the subjects were not hGH deficient at the time of study.

Root et al. [18] also studied hGH secretion in children with the FAS. All 4 studied were of low birth weight indicating the presence of intrauterine growth retardation. Microcephaly and other features of prenatal ethanol toxicity were present in all. The patients were somewhat older (9.9–14.5 years) than those studied by *Tze* et al. [26] and were all born to the same mother. All 4 had normal hGH responses to insulin-induced hypoglycemia but their peak levels ranged from 8.2 to 24.2 ng/ml – similar to those found in normals.

The discussion in *Tze* et al. [26] does not clearly distinguish between an inhibitory effect on hGH secretion when ethanol is present in the maternal and fetal circulation and permanent deficiency persisting in the infant long after exposure has been terminated by parturition. Infants with the FAS evidently do not have hGH deficiency months or years after birth. However, hGH levels in mother or infant have never been measured during actual exposure. Since ethanol inhibits GH release under other circumstances, it is likely that it also does so in the pregnant woman and fetus.

Some infants with severe intrauterine growth retardation continue to show impaired growth postnatally even though the factor constraining growth is no longer present. Hence, the biochemical abnormalities restraining growth in the alcohol-exposed fetus may no longer be detectable months or years later. There is a substantial body of animal investigation indicating that maternal GH is important for fetal outcome particularly when the mother is nutritionally compromised [19, 27].

The rGH deficiency produced in the rat by perphenazine is sufficient to produce growth retardation, indicating that inhibition of GH secretion by a drug can be biologically significant [17]. Since GH does not cross the placenta, maternal and fetal secretion are autonomous. It has frequently been stated that hGH is thought to be inactive in the fetus, largely because fetal blood levels of somatomedin are low. Recently, it has been reported that somatomedin synthesized within the tissue itself may be important in stimulation of fetal growth [4]. Thus, it is likely that fetal blood levels of somatomedin do not fully reflect the biological role of this growth factor. In view of evidence that both maternal and fetal GH secretion may be important in promotion of fetal growth, the possibility that inhibition of hGH secretion by ethanol is involved in the pathogenesis of the FAS would appear to merit further investigation.

Conclusions

The preponderance of evidence in human studies indicates that ethanol inhibits hGH release. Our studies in the rat demonstrate that high, but not low, doses of ethanol stop rGH secretion for at least 3 h. There does not seem to be any time period after any ethanol dose employed when rGH levels are higher. While the single human study reporting an increase early after ethanol [3] cannot be discounted, it is clear that the major effect of ethanol in high doses is to inhibit ethanol release. If changes in hGH secretion play any role in the pathology of chronic alcoholism, then it is deficiency rather than excess which is involved.

Summary

Growth hormone (GH) regulates not only somatic growth but also carbohydrate, protein and lipid metabolism. Altered secretory states resulting from alcohol use could have pathogenetic significance. The pattern of spontaneous GH secretion in man is reviewed and previously published human studies on ethanol and GH examined to develop a coherent formulation of the nature of ethanol effects on GH secretion. The weight of the evidence suggests that ethanol suppresses GH secretion. Studies using the rat indicates that the dose-response relationship is of a threshold nature, with doses of 3 g/kg or greater abolishing spontaneous secretion. The possible role of diminished GH secretion in the pathogenesis of the fetal alcohol syndrome is discussed.

References

1 Arky, R.A.; Freinkel, N.: The response of plasma human growth hormone to insulin and ethanol-induced hypoglycemia in two patients with 'isolated adrenocorticotropic defect'. Metabolism *13:* 547–550 (1964).
2 Bansal, S.; Lee, L.; Woolf, P.D.: Dopaminergic modulation of arginine mediated growth hormone and prolactin release in man. Endocrine Annu. Meet. Abstr. No. 479 (1980).
3 Bellet, S.; Yoshimine, N.; Decastro, O.A.P.; Roman, L.; Parmar, S.S.; Sandberg, H.: Effects of alcohol ingestion on growth hormone levels: their relation to 11-hydroxycorticoid levels and serum FFA. Metabolism *20:* 762–769 (1971).
3a Ellis, F.W.: Effect of ethanol on plasme corticosterone levels. J. Pharmac. exp. Ther. *153:* 121–127 (1966).
4 D'Ercole, A.J.; Applewhite, G.T.; Underwood, L.E.: Evidence that somatomedin is synthesized by multiple tissues in the fetus. Endocrine Annu. Meet. Abstr. p. 220 (1979).
5 Durand, D.; Martin, J.B.; Brazeau, P.: Evidence for a role of α-adrenergic mechanisms in regulation of episodic growth hormone secretion in the rat. Endocrinology *100:* 722–728 (1977).
6 Finkelstein, J.W.; Roffwarg, H.P.; Boyar, R.M.; Kream, J.; Hellman, L.: Age-related changes in the twenty-four hour spontaneous secretion of growth hormone. J. clin. Endocr. Metab. *35:* 665–670 (1972).
7 Ganda, O.P.; Sawin, C.T.; Iber, F.; Glennon, J.A.; Mitchell, M.L.: Transient suppression of growth hormone secretion after chronic ethanol intake. Alcoholism *2:* 297–299 (1978).
8 Leppaluoto, J.; Rapeli, R.; Varis, R.; Ranta, T.: Secretion of anterior pituitary hormones in man: effects of ethyl alcohol. Acta physiol. scand. *95:* 400–406 (1975).
9 Martin, J.B.; Reichlin, S.; Brown, G.M.: Regulation of growth hormone secretion and its disorders. Clinical neuroendocrinology (Davis, Philadelphia 1977).
10 Othmer, E.; Goodwin, D.; Levine, W.; Malarkey, W.; Freemon, F.; Halikas, J.; Daughaday, W.: Sleep-related growth hormone secretion in alcoholics. Clin. Res. *20:* 726 (1972).
11 Ouellette, E.M.; Rosett, H.L.; Rosman, N.P.; Weiner, L.: Adverse effects on offspring of maternal alcohol abuse during pregnancy. New. Engl. J. Med. *297:* 528–530 (1977).
12 Parker, D.C.; Rossman, G.; Kripke, D.F.; Gibson, W.; Wilson, K.: Rhythmicities in human growth hormone concentrations in plasma; in Krieger, *Endocrine rhythms* (Raven Press, New York 1979).

13   Plotnick, L.P.; Thompson, R.G.; Kowarski, A.; deLacerda, L.; Migeon, C.; Blizzard, R.M.: Circadian variation of integrated concentration of growth hormone in children and adults. J. clin. Endocr. Metab. *40:* 240–247 (1975).
14   Priem, H.A.; Shanley, B.C.; Malan, C.: Effect of alcohol administration on plasma growth hormone response to insulin-induced hypoglycemia. Mebabolism *25:* 397–403 (1976).
15   Redmond, G.P.: Effect of ethanol on endogenous rhythms of growth hormone secretion. Alcoholism *4:* 50–56 (1980).
16   Redmond, G.P.: Effect of perphenazine on secretory patterns of growth hormone in the rat. Neuroendocrinology *30:* 243–248 (1980).
17   Redmond, G.P.; Hirshman, M.F.: Post-natal growth retardation in rats produced by the phenothiazine perphenazine. Pediat. Pharmacol. *1:* 153–160 (1980).
18   Root, A.W.; Reiter, E.O.; Andriola, M.; Duckett, G.: Hypothalamic-pituitary function in the fetal alcohol syndrome. J. Pediat. *87:* 585–588 (1975).
19   Sara, V.R.; Lazarus, L.; Stuart, M.C.; King, T.: Fetal brain growth: selective action by growth hormone. Science *186:* 446–447 (1974).
20   Steiner, R.A.; Stewart, J.K.; Barber, J.; Koerker, D.; Goodner, C.J.; Brown, A.; Illner, P.; Gale, C.C.: Somatostatin: a physiological role in the regulation of growth hormone secretion in the adolescent male baboon. Endocrinology *102:* 1587–1594 (1978).
21   Stewart, J.K.; Koerker, D.J.; Goodner, C.J.; Gale, C.C.; Steiner, R.A.: Neural and metabolic regulation of spontaneous growth hormone secretion in the primate. Endocrine Annu. Meet. Abstr. No. 477 (1980).
22   Streissguth, A.P.; Landesman-Dwyer, S.; Martin, J.C.; Smith, D.W.: Teratogenic effects of alcohol in humans and laboratory animals. Science *209:* 353–361 (1980).
23   Tamburrano, J.; Tamburrano, S.; Gambardella, S.; Andreani, D.: Effects of alcohol on growth hormone secretion in acromegaly. J. clin. Endocr. Metab. *42:* 193–196 (1976).
24   Tannenbaum, G.S.; Martin, J.B.: Evidence for an endogenous ultradian rhythm governing growth hormone secretion in the rat. Endocrinology *98:* 562–570 (1976).
25   Toro, G.; Kolodny, R.C.; Jacobs, L.S.; Masters, W.H.; Daughaday, W.H.: Failure of alcohol to alter pituitary and target organ hormone levels. Clin. Res. *21:* 505 (1973).
26   Tze, W.J.; Friesen, H.G.; MacLeod, P.M.: Growth hormone response in fetal alcohol syndrome. Archs Dis. Childh. *51:* 703–706 (1976).
27   Zamenhof, S.; Vanmarthens, E.; Grauel, L.: Prenatal cerebral development: effect of restricted diet, reversal by growth hormone. Science *174:* 954–955 (1971).

Dr. G.P. Redmond, Department of Pharmacology, University of Vermont College of Medicine, Burlington, VT 05405 (USA)

Alcoholism and Reproduction

M. Wayne Heine

Department of Obstetrics and Gynecology, Texas Tech University, Health Sciences Center, School of Medicine, Lubbock, Tex., USA

Introduction

Diligent students of history know that alcohol and reproduction are incompatible. Judges 13:4 tells us 'now, therefore beware, I pray thee drink not wine nor strong drink.' Later in Judges 13:7 the Bible tells us, 'But he said unto me, behold thou shalt conceive and bury son and not drink no wine nor strong drink, neither eat any unclean thing for the child shall be a Nazarite unto God from the womb to the day of his death.' The ancient Greeks of Sparta and Carthage forbade the use of alcohol on the wedding day. Greek mythology tells us that alcohol caused deformity of the god, Volcanus. More recently, the English parliament in the 1830s became concerned about the 'puny children' from the gin epidemic that occurred with decreased liquor costs. Essentially, history tells us reproduction and alcohol do not mix.

Alcohol and its Effect on Reproduction

Male Reproduction

Male reproductive function is affected in many ways by alcohol. This will be discussed in other sections of these proceedings. Alcohol may have an immediate effect on the male with diminished sexual performance. There are also a number of long-term effects of alcohol on the reproductive system of the male. As liver function declines there is a decreased clearance of estrogens, which in turn, have far-reaching systemic effects. *Lox* et al. [10], in our laboratory, found in a study of 14 males in a detoxification center, that testosterone values were below the normal limits in 4 subjects which is in agreement with an initial report [16] with diminished testosterone values and elevated estrogen levels, the male will often have testicular atrophy, gynecomastia, libido decrease and fertility decrease. By the time the male has reached this state in the disease process, little can be done to improve sexual performance or fertility. Clomiphene citrate therapy has been used on a short-term basis to improve both potency and fertility. At a daily dose of 50 mg of clomiphene citrate, *Bjork* [1] noticed an increase in LH and testosterone. Using 100 mg daily of clomiphene, both FSH and potency were increased. Clomiphene citrate must be utilized on a short-term basis, however, long-term usage of clomiphene, since it is a very weak estrogen, may further decrease testicular size and fertility. Clomiphene-mediated effects on the liver are not presently known.

Female Reproduction

The incidence of alcoholism in females appears to be increasing. It is presently estimated that the incidence among the adult female population of an alcohol problem is somewhere between 3 and 9% [12]. Studies concerning alcohol and its effect on the reproductive cycle are very limited and only of recent origin. This may be related to our unawareness of the high incidence of alcoholism in the female population. One has a great deal of difficulty quantitating the amount of alcohol ingested by many women. For example, it is much easier for a female to be a 'closet drinker'. Females are more apt to be 'junk drinkers', sipping sherry or wine throughout the day, then having a drink with their husbands in the evening before, and possibly even after, dinner. It is also easier for women to hide inadequacies of sexual function since impotence is a common symptom in males which is much more difficult to hide.

Many of us who care for females have not been aware of the problem of alcoholism and have been less than diligent in obtaining a drinking history. When a woman stays home and is alone most of the day she may consume large quantities of alcohol in relative privacy with the disease process not being picked up until it is well advanced. As women have entered the work forces and physicians are more aware of the problems of alcoholism, we may be recognizing a problem that has been there all along.

Pharmacokinetics. Women handle alcohol metabolically different than men. *Jones* et al. [5] showed that alcohol absorption is lowest early in the cycle and markedly increased late in the cycle. Therefore, females have more rapid absorption and higher alcohol levels on ingestion of equal amounts of alcohol than males. This may explain why females do less well in cognitive tasks than males when given comparable amounts of alcohol. *Jones and Pardes* [5] showed that there is some delay in alcohol absorption in women taking the birth control pill but the plasma alcohol levels are prolonged for at least an hour longer. The fact that women have more rapid absorption of alcohol than males may explain why they develop liver diseases more rapidly than their male counterparts.

Alcohol-Drug Interaction and Ovulation. Alcohol appears to be one psychotropic drug that fooled mother nature. The majority of drugs that exert their action on the central nervous system will block ovulation. Alcohol usually blocks the estrus cycle in experimental animals but in the adult female, it is rare that a woman will be anovulatory on the basis of alcohol alone. *Ryback* [15] did report 2 cases of females with amenorrhea during the acute phase of their alcoholism with restoration of menstrual cycles following detoxification. This is an unusual event with alcohol itself. Since many females take multiple over-the-counter or prescription drugs as well as alcohol, many of these central nervous acting agents may block ovulation, particularly in advanced alcoholism where there is a poor metabolic clearance of estrogen, and the female then may become anovulatory. Therefore, alcoholic females, except in the terminal state of their disease process, do have potential to conceive.

Although it is difficult to translate animal studies into human application we do obtain some information from animal research in the area of alcohol and reproduction. It appears that during the acute ingestion of alcohol, sperm migration may be arrested and there may be a direct toxic effect of alcohol on sperm in the fallopian tubes of experimental animals [16]. In our laboratory we found either alteration or blockage of estrus. Long-term usage among experimental animals usually blocks estrus cycle [7]. We found an increase in prolactin levels in experimental animals [11]. If applied to the humans, this could explain a shortened luteal phase and an increasing incidence of abortion.

Alcohol Consumption during Pregnancy of the Alcoholic Women. Little and Streissguth [9] have shown that regular drinkers may decrease their consumption during pregnancy as much as 50%. Although binge drinkers may show a decrease in the amount of alcohol consumed in pregnancy, the number of binges may increase by 20%. So there is an inherent tendency to decrease the alcohol consumption in pregnancy. They suddenly become concerned about the physiologic effect of alcohol on themselves and the baby or there may be less social pressure on them and less need for alcohol. Experimental studies, i.e., particularly those related to alcohol-addicted pigs and monkeys, indicated that those who conceive will automatically decrease their consumption of alcohol in pregnancy [11]. The alcoholic female, when pregnant, will respond differently to alcohol. Since the GI emptying time is longer in pregnancy the levels of alcohol will be lower but will last longer. It appears that high levels, not the length of exposure, is the toxic stimulus to the fetus. Therefore, mother nature compensated for this phenomenon of letting the female ovulate with alcohol problems by protecting the fetus by causing slower absorption of alcohol. Women will show a greater tolerance in pregnancy to alcohol.

Alcohol and Pregnancy. One of the true emergencies which occur in the alcoholic pregnant female is the problem of ketoacidosis. *Dillon* et al. [2] first described this problem in 1940. *Podratz* [14] reported such a case, with mild liver dysfunction and increased alcohol history, having acute ketoacidosis. He states this is an acute emergency. The patients will have ketosis, hyperpnea, dehydration, but will have a normal blood glucose, and absent glucosuria, which will help the physician differentiate this from diabetic ketoacidosis. We know from diabetic ketoacidosis that this is a medical emergency associated with both maternal and fetal mortality. These patients should be admitted immediately, treated with IVs, electrolyte restoration, and given glucose. Some physicians recommend utilization of insulin to increase glucose uptake.

In the past it was felt that alcoholism was associated with premature labor and delivery with a reported 15% incidence of prematurity. More recent studies, such as *Kaminiski* et al. [6], showed these women went to term but there was intrauterine growth retardation, and the infants were term in length of gestation but were premature in weight. Thus, it appears that this is not prematurity but intrauterine growth retardation. *Little and Streissguth* [9] reported that alcohol consumed daily but stopped prior to pregnancy was associated with 91 g weight decrease in the infants when compared to controls. When alcohol was consumed in the third trimester the decrease in weight was 160 g. So it appears that low alcohol ingestion or discontinuing the alcohol ingestion just prior to pregnancy is associated with decrease in weight as well as the fetal alcohol syndrome. The incidence of small-for-date deliveries in *Kaminiski* et al. [6] series among heavy drinkers was 4.8%, and among light drinkers, 2.5%. Other problems in pregnancy noted in the study of *Kaminiski* et al. [6] was an increased incidence of stillbirths. This report, from France, showed heavy drinkers having a 2.6% incidence of stillbirths among heavy drinkers and a 1% incidence among light drinkers. Of interest was that heavy beer drinkers had a stillbirth rate of 3.8% [6]. There was also an increased incidence of abruptio placentae among these women with 3 out of 4 being among the beer drinkers. Since this study was performed in France the majority of these women consumed excessive amounts of wine and/or beer. This may be of interest as the drinking habits of the younger generation in this country have turned away from hard liquor and are consuming increasing amounts of wine and beer.

Alcohol and the Fetus. It was not until the study of *Lemoine* et al. [8] in 1968 and the article of *Jones* et al. [4] in 1973 that we began to crystalize the far-reaching effects of alcohol on the fetus. It was in these reports that the fetal alcohol syndrome was described. This problem will be discussed elsewhere in the present proceedings, but briefly, the fetal alcohol syndrome consists of facial anomalies, decreased prenatal brain growth, mental deficiency, and multiple other anomalies. *Quelette* et al. [13] found 32% of the infants of heavy drinkers to have anomalies. 14% of the infants of moderate drinkers had anomalies, and 9% of the light drinkers, which is close to that reported in most general populations. It is difficult to separate the toxic effect of alcohol on the fetus from the many other problems that alcoholic women may have which may be associated with increased incidence of anomalies. Since the immune mechanism in pregnancy is depressed, and also the immune mechanism among alcoholics is depressed, the pregnant alcoholic female has an increased susceptibility to various infections such as viral infections which may have a teratogenetic effect by themselves or in synergism with alcohol. Most alcoholic females are also malnourished, and the majority of the abnormalities described with alcohol, have also been associated with malnutrition. Most alcoholic individuals have a history of ingestion of multiple drugs and/or tobacco usage. We know the tobacco is associated with decreased placental blood flow, and small-for-date infants [3]. If there is some liver damage drugs taken by the female will have decreased clearance; therefore, higher and longer levels may increase the risk of fetal abnormalities. In another survey of 231 middle-class pregnant females where it was shown that 95% of these females took over-the-counter drugs, 68% used alcohol and 23% smoked. An alcoholic female may take an over-the-counter drug which would, in a normal individual, be cleared and cause no difficulties. In the alcoholic female, it may potentially be lethal or toxic to the developing fetus.

The fetal alcohol syndrome has been reported to occur as high as 2/1,000 deliveries. It has not been that high in my experience. Of major concern is the effect of alcohol on the brain growth of a developing fetus and the number of mentally handicapped babies that may be born to mothers ingesting excessive amounts of alcohol.

Conclusion

What is the safe level of alcohol? This is an extremely difficult question. Since human placentation is different from most of the animal models we have used in the study of alcoholism, it is difficult to translate our animal studies to human application. We do know, from retrospective studies, that ingestion of from 2 to 4 oz of alcohol is associated with an increased incidence of fetal abnormalities. Therefore, consumption of no alcohol is ideal. From animal studies, and our own patient experience, we know that alcoholic females are often motivated to seek treatment during pregnancy. Therefore, it behooves those of us who take care of the pregnant female, be it family practitioner or obstetrician-gynecologist, to beware of the problem of alcohol and pregnancy. First, we need to obtain a drinking history. We should be alert for the signs, and aware of the history of the alcoholic female. If there is a family history of depression or alcoholism in the family the woman has an increased risk of developing alcoholism or has a greater chance of being an alcoholic. If we find a patient who misses many appointments or is undependable or, perhaps, during our physical examination we see multiple bruises, we should be alert for the problems of alcoholism. Since the alcoholic is great at denials, pregnancy may be a splendid opportunity for patient education about the problem of alcohol, not only on her own body, but on the developing fetus. By being more diligent in our observation and history taking, we may be able to prevent literally millions of children from becoming mentally handicapped. We may also be able to put many women on treatment programs and allow them to return as productive members of society. The treatment of alcoholic females, as well as their mentally handicapped children, and the home environment in which they are raised, costs our society billions of dollars. These few minutes of extra history taking may well be the most cost-effective health care that we can provide.

Summary

A brief overview on the reproductive capacities of both men and women in alcoholism is presented. A historical evaluation indicates a resurgence of interest in this area. The effect of chronic alcohol consumption on both male fertility and potency is reported in conjunction with alcohol-mediated effects on the female subject. Emphasis is placed on pharmacokinetics, metabolism and drinking behavior of the alcoholic female. The adverse actions of some therapeutic drugs and chronic alcohol consumption is discussed in relationship to fetal alcohol syndrome and the accompanied mental and somatic abnormalities.

References

1 Bjork, J. T. et al.: Clomiphene citrate therapy in a patient with Laennec's cirrhosis. Gastroenterology *12:* 1308–1311 (1977).
2 Dillon, E. S.; Dyer, W. W.; Smelo, L. S.: Ketone acidosis in non-diabetic adults. Med. Clins N. Am. *24:* 1813 (1940).
3 Hasse, H. E.; Waldmann, H.; Schoenhoefer, P. S., et al.: Behandlungsprinzipien bei jugendlichen Drogenkonsumenten. Rhein. Ärztebl. *23:* 795 (1971).
4 Jones, K. L.; Smith, D. W.; Ulleland, C. W., et al.: Pattern of malformation in offspring of chronic alcoholic mothers. Lancet *1:* 1267–1271 (1973).
5 Jones, B. M.; Jones, M. K.; Pardes, A.: Oral contraceptives and ethanol metabolism. Meet. XVth Reunions Anual de la Sociedad Mexicana de Nutricion y Endocrinologia, Acapulco 1975.
6 Kaminiski, M.; Rumea Rouguette, C.; Schwartz, J.: Consommation d'alcool chez les femmes enceintes et issue de la grossesse. Revue Epidém. Santé publ. *24:* 27–40 (1976).
7 Kieffer, J. D.; Ketchel, M. A.: Blockade of ovulation in the rat by ethanol. Acta endocr., Copenh. *65:* 117 (1970).
8 Lemoine, P.; Hakonsseam, H.; Borteyru, J. P., et al.: Les enfants de parents alcooliques: anomalies observées à propos de 127 cas. Ouest méd. *25:* 476–487 (1968).
9 Little, R. E.; Streissguth, A. P.: Drinking during pregnancy in alcoholic women. Alcoholism: Clin. exp. Res. *2:* 179–182 (1978).
10 Lox, C. D.; Peddicord, O.; Heine, M. W.; Messiha, F. S.: The influence of chronic

long-term and alcohol abuse on testosterone secretion in men and rats. Proc. West. Pharmacol. Soc. *21:* 299–302 (1978).
11 Lox, C.D.; Messiha, F.S.; Heine, M.W.: The relationship between chronic ethanol intake and oral contraceptives on reproductive function in the female rat (to be published, 1981).
12 Morse, R.M.; Hurt, R.D.: Screening for alcoholism. J. Am. med. Ass. *242:* 2688–2690 (1979).
13 Ouelette, E.M.; Rosett, H.L.; Rosman, N.P.; Weiner, L.: Adverse effects on offspring of maternal alcohol abuse during pregnancy. New Engl. J. Med. *297:* 528–530 (1977).
14 Podratz, K.C.: Alcoholic ketoacidosis in pregnancy. Obstet. Gynec. *52:* 54–57 (1978).
15 Ryback, R.S.: Chronic alcohol consumption and menstruation. J. Am. med. Ass. *238:* 2143 (1977).
16 Sherma, S.C.; Chandhury, R.R.: Studies on the effect of ethanol. J. med. Res. *58:* 505 (1970).
17 Van Thiel, D.H.; Lester, R.: Sex and alcohol. New Engl. J. Med. *291:* 25–53 (1974).

Dr. M.W. Heine, Department of Obstetrics and Gynecology, Texas Tech University, Health Science Center, School of Medicine, Lubbock, TX 79430 (USA)

Fetal Alcohol Syndrome: Neurochemical and Endocrinological Abnormalities

Pushpa V. Thadani

Veterans Administration Medical Center, Washington, D.C., USA

The potential teratogenic effects of alcohol on offspring have been suspected since ancient times. In 1899, *Sullivan* [73] recorded the damaging effects of maternal intoxication on the fetus but this early work remained unnoticed by the medical research community until recently.

In recent years, a particular interest has developed in alcohol as a teratogen because it is widely used by women and also because a wide-range of adverse effects have been seen in offspring exposed in utero to drugs of abuse or other teratogens (e.g. thalidomide). Alcohol, because of its molecular weight, water-lipid solubility and ionic nature, crosses the placental barrier easily even late in pregnancy [29, 88] and thus, can effect the development of the fetus directly.

Fetal Alcohol Syndrome

In 1968, *Lemoine* et al. [44] reported that infants born of alcoholic mothers had abnormal facial characteristics, growth deficiencies and psychomotor disturbances. Five years later, *Jones and Smith* [35] independently described a similiar pattern of malformation in 8 unrelated children born to chronically alcoholic mothers. These children displayed a constellation of particular facial features, growth deficiencies and mental retardation that *Jones and Smith* [35] termed the 'fetal alcohol syndrome' (FAS). Since then, many case reports of FAS have been published from around the world [11, 21, 39, 49]. The perinatal mortality of infants with FAS is high, and those who survive show manifestation of neurological difficulties and mental deficiencies of varying severity [11, 21, 31, 36, 37, 71].

In recent years, evidence has accumulated which indicates that even two drinks of alcohol taken per day during pregnancy could effect the development of offspring [31, 35, 36, 45, 46, 52, 71]. In addition, these studies also indicate that the deleterious effects of alcohol may be dose-dependent as the effects in offspring can range from lowered birth weight and functional deficits to FAS. Since it is difficult to determine what the safe level of alcohol is for a pregnant woman, animal studies have been employed to address this and other critical issues of FAS.

Animal Models

Teratogenic and Behavioral Effects of Ethanol

Earlier animal studies have reported that maternal ethanol intake caused damaging effects on the development of the fetus [51]; however, the animal models directed specifically to study FAS are relatively recent [19, 24, 28]. Animal models are particularly important because a number of factors, such as dose, length of exposure, nutrition by pair-fed controls and the postnatal environment can be controlled.

In several species of animals, studies have shown that maternal ethanol ingestion during pregnancy increases the fetal mortality rate and decreases the litter size and weight of the fetus and offspring [1, 3, 4, 41, 55, 65, 66, 86]. Morphological abnormalities have also been found in several tissues of the neonates exposed to ethanol either in utero or between 24 and 96 h after birth [8, 12, 13, 20, 25, 34, 56]. Exposure to ethanol during gestation causes delayed onset of myelination and lamination in the cerebral cortex of the offspring [34]. Several investigators have shown that postnatal exposure of neonatal rats to ethanol caused impaired growth and destruction of neural cells in the cerebellum [8, 12, 13, 25].

Behaviorally, maternal exposure to alcohol influences emotionality and learning ability at maturity in offspring [2, 9, 14, 16, 23, 47, 54, 62, 63, 74, 87]. In addition, ethanol-exposed animals show hyperactivity [15, 47] when measured by a variety of methods. It should be pointed out that some of these findings replicate clinical reports seen in FAS children.

Neurochemical and Hormonal Effects of Ethanol

Biochemical, hormonal and neural abnormalities can occur in the offspring exposed to a drug in utero in the absence of gross or classical teratological effects. These abnormalities at the cellular or subcellular level may have profound effects on the offspring's function. Studies have shown that ethanol given during gestation causes biochemical (including neurotransmitters) and functional changes in offspring in the absence of gross physical changes [18, 30, 32, 33, 40, 42, 43, 59, 60, 80, 82].

Several investigators have reported that both acute and chronic administration of ethanol to pregnant animals caused alteration in protein synthesis in fetal and neonatal brain and liver; however, the type of change in protein synthesis seen in these tissues depended on factors like the length of exposure to ethanol, maternal blood ethanol level and the postnatal age of the neonates [32, 33, 40, 59, 60]. These studies also show that maternal ethanol intake produced changes in total DNA and RNA in fetal and neonatal tissue [32, 60]. Maternal ethanol ingestion did not induce the activity of enzyme alcohol dehydrogenase in young or adult offspring [27, 68]. Similar results have been reported in adult animals given alcohol chronically.

Both biochemical and morphological alterations in myelination of the central nervous system (CNS) have been reported [30, 34]. In a biochemical study, an excess of the chemically and morphologically immature heavy myelin was observed in ethanol-exposed pups suggesting that the myelination of the CNS is abnormal after ethanol exposure [26, 30]. This study also reports that the severity of this adverse effect on CNS myelination depended on the length of maternal ethanol intake. In morphological studies, intial retardation in cortical development has been shown in pups exposed to ethanol during gestation suggesting that the abnormalities in function of the cerebral cortex may develop in later life [34].

In the past decade evidence has accumulated to suggest that neurotransmitter systems in the developing brain and autonomic nervous system are particularly sensitive to drugs which exert their action on neural function. Previous studies have shown that in adult animals, administration of acute and chronic ethanol causes alteration in brain ^{3}H-norepinephrine disposition [79, 84]. This prompted us to investigate the effects of chronic maternal ethanol ingestion on the development of both central and peripheral adrenergic system in the offspring.

In pups exposed continuously to ethanol both pre- and postnatally, significant increases in synaptosomal tyramine uptake and octopamine synthesis were observed suggesting that this exposure of developing rats to ethanol increases either the number or function of noradrenergic synapses [80]. Since these elevations disappeared by 24 days of age it suggests that these alterations in noradrenergic synapses are not permanent in ethanol-exposed pups.

To determine the extent to which ethanol-induced alterations in the development pattern of synaptic function could be prevented by termination of exposure, animals were withdrawn at different postnatal ages [80]. Withdrawal appeared to enhance the tendencies toward accelerated synaptogenesis suggesting that postnatal withdrawal from ethanol does not prevent and may actually worsen the alterations in central noradrenergic development.

In contrast to the effects of continuous ethanol on synaptosomal uptake, little change was observed in tyrosine hydroxylase activity and brain weights suggesting that the observed alterations in noradrenergic development are not secondary to nonspecific nutritional or growth deficiencies. *Branchey and Friedhoff* [18] report an increase in caudate nucleus tyrosine hydroxylase in pups exposed to ethanol prenatally but withdrawn at birth. The lack of any change after continuous ethanol administration suggests that the enzyme alteration may result from the stress of withdrawal rather than from the direct effects of ethanol. Additionally, studies done in whole brain may mask regional changes.

In pups exposed to ethanol both pre- and postnatally, a deficiency was observed both in adrenal catecholamines stores and in dopamine β-hydroxylase activity, the enzyme associated with storage vesicles, throughout postnatal development. These data suggest that maternal ethanol ingestion causes a significant reduction of adrenal catecholamines stores, partially due to the effects on storage vesicle development [43]. These actions of ethanol on adrenals were fully reversible upon withdrawal within the 1st week of postnatal life. This is in contrast to the actions seen on the central noradrenergic system [80].

These findings with neurotransmitters raised the question whether these are primary effects of ethanol or these changes are secondary to a more general interference with the cellular development of neural tissues. One of the biochemical tools used in recent years to evaluate general growth and development is the activity of enzyme ornithine decarboxylase (ODC). ODC catalyzes the conversion of ornithine to putrescine, the first and probable rate-limiting step in polyamine biosynthesis [64, 75]. It has been suggested that polyamines may play a regulating role in protein synthesis [5, 10, 57, 58].

During fetal and neonatal brain development, the highest ODC and polyamine levels are paralleled by the period of most rapid cellular growth and replication [6, 53] and at these times, the maturational pattern of ODC activity appears to be particularly sensitive to hormonal or drug influence [7, 17]. These data suggest that perturbations in this pattern may represent an early index of disturbed development of the CNS.

In pups exposed continuously to ethanol, the developmental pattern of ODC activity in *brain* was altered in a complex fashion: an initial decline in activity was followed by a rebound to normal or supranormal levels before declining again [81, 82]. Maternal ethanol ingestion also caused alteration in *heart* ODC activity but the changes in developmental pattern were different from brain. In heart, an initial increase in activity was followed by a premature decline to low activity characteristic of adult heart tissue [81, 82].

To determine the extent to which the ethanol-induced abnormality in ODC development could be prevented by postnatal termination of ethanol exposure, animals were transferred to control mothers at various postnatal ages. Withdrawal from ethanol at different postnatal ages produced alterations in brain and heart ODC activity that were different from those seen in pups exposed to ethanol continuously. Withdrawal at birth produced the most dramatic effect on heart weight and its ODC acitivity [81, 82]. These data suggest that the biochemical changes are dependent on the duration of exposure as well as time at which withdrawal is initiated. In addition, these data also indicate that either pre- or perinatal ethanol exposure significantly alters polyamine metabolism and growth in developing brain and heart. The effects of maternal ethanol ingestion on the development of ODC may play a role in alterations seen in protein and nucleic acid synthesis in these tissues [32, 33, 40, 59, 60].

Recent studies have shown that the activity of ODC in various tissues may be regulated by growth hormone (GH) [67]. Earlier investigations have shown that in adults and developing rats, GH secretion is regulated by the biogenic amines [22, 48, 50, 69, 70, 72]. In animals, ethanol causes alterations in brain ODC activity and development of the noradrenergic system and *in infants* born of alcoholic mothers, hypersecretion of GH occurs after insulin-induced hypoglycemia and after

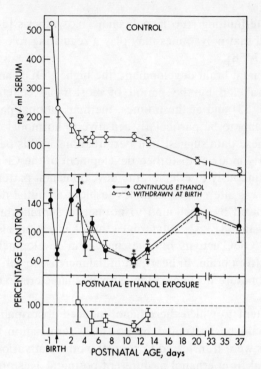

Fig. 1. Serum growth hormone levels in developing rats. Continuous ethanol denotes maternal ingestion from the 13th day of gestation and continued for the duration of the experiment. Pups born of ethanol-treated mothers were transferred to control mothers at birth (withdrawn group) and control pups were reared by ethanol-treated, started from 13th day of gestation mothers (postnatal group) as indicated. Points and bars represent means ±-SE of 5–15 determinations at each age; asterisks denote significant differences versus control (at least $p < 0.05$ by unpaired t test).

arginine infusion [85]. These observations led us to investigate the effects of chronic maternal ethanol ingestion on GH secretion in the offspring.

In neonates exposed to ethanol continuously, serum GH levels were increased initially and then declined below normal until weaning suggesting that there may be an alteration in the development of hypothalamic regulation of GH secretion during the early neonatal period (fig. 1) [83]. These abnormal levels of serum GH seen in neonates exposed to ethanol throughout development were not prevented by termination of ethanol exposure at birth (fig. 1).

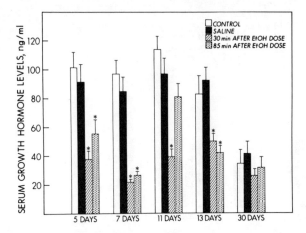

Fig. 2. Effect of acute ethanol dose (2 g/kg) on serum growth hormone levels in developing rats 30 or 85 min after ethanol injection. Bars represent means ±-SE of 12–24 determinations at each age; asterisks denote significant differences versus saline/untreated control (at least $p < 0.05$ by unpaired t test).

To determine whether a significant effect of ethanol could be achieved solely by postnatal exposure, animals born to control mothers were transferred to ethanol-treated mothers (treated from 13th day of gestation); alternately, control neonates were administered a single dose of ethanol, 2 g/kg. Postnatal ethanol exposure in the former group caused a decrease in serum GH concentration starting from day 5 (fig. 1). Similarly, a single dose of ethanol given to the developing rats (in the later group) produced a marked decrease in serum GH level within 30 min (except in 30-day olds) indicating that ethanol can interfere acutely with GH secretion (fig. 2) [83]. However, it is not clear from these data whether this action of ethanol is due to a direct effect on the hypothalamic-pituitary axis regulation of GH secretion or whether it is secondary to a 'nonspecific' stress effect produced by ethanol.

Acute ethanol also caused decreases in brain and heart ODC activity in developing rats (fig. 3, 4) [83]. These decreases in activity correlate both in time and effect with the suppression of GH levels in the serum. The present data would be consistent with the hypothesis that decreased secretion of GH might mediate the alcohol-induced alteration in tissue ODC activity.

As a single dose of ethanol caused marked alterations in secretion of GH when administered postnatally to rat pups, it was important to in-

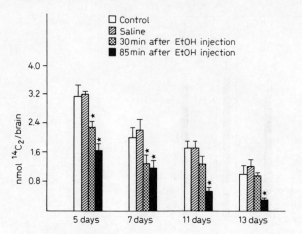

Fig. 3. Effect of acute ethanol dose (2 g/kg) on brain ornithine decarboxylase activity in developing rats 30 or 85 min after ethanol injection. Bars represent means ± -SE of 6–12 determinations at each age; asterisks denote significant differences versus untreated control/saline (at least $p < 0.05$ by unpaired t test).

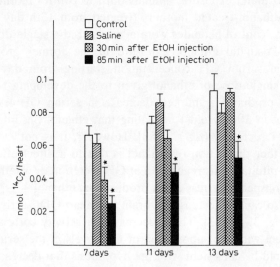

Fig. 4. Effect of acute ethanol dose (2 g/kg) on heart ornithine decarboxylase activity in developing rats 30 or 85 min after ethanol injection. Bars represent means ± -SE of 5–12 determinations at each age; asterisks denote significant difference versus saline/untreated control (at least $p < 0.05$ by unpaired t test).

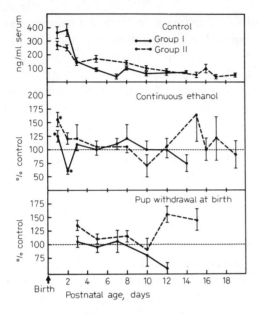

Fig. 5. Serum growth hormone levels in developing rats. In group I, rats were exposed to the *ethanol dose* (4 g/kg) *daily* from the 13th day of gestation and in group II, from the 18th day of gestation. In the withdrawal experiments, pups were transferred from ethanol-treated mothers to control mothers at birth. Points and bars represent means ±-SE of 6–12 determinations at each age; asterisks denote significant differences (at least $p < 0.05$) versus control by unpaired t test.

vestigate further the actions of a daily dose of ethanol on GH secretion in the offspring. Pregnant rats were given an oral dose of ethanol, 4 g/kg, daily either from the 13th or 18th day of gestation and thereafter. Controls received orally an isocaloric dose of sucrose in saline.

In neonates exposed to a daily dose of ethanol from the 13th day of gestation and thereafter, serum GH concentrations were increased initially and then declined to subnormal levels before returning to normal levels by day 4 (fig. 5). However, the secretion of GH in pups exposed to ethanol from the 18th day of gestation (group II) remained elevated until day 8 suggesting that exposure to a moderate dose of ethanol daily during development causes alteration in GH secretion. Alterations in the development pattern of brain and heart ODC activity were also observed [77, 78].

To determine whether termination of ethanol exposure would prevent abnormal GH secretion, pups at birth were transferred to control mothers (withdrawal at birth). Withdrawal at birth did prevent abnormal GH secretion in pups exposed to ethanol daily from the 13th day of gestation (group I, fig. 5) but not in pups exposed to ethanol from the 18th day of gestation (group II, paired t test, $p < 0.05$ through 8 days of age). These results indicate that the day of gestation on which exposure to ethanol was initiated may be critical to the type of alteration seen in GH secretion. However, from these data it is not clear whether the action of ethanol is due to a direct effect on the hypothalamic-pituitary axis regulation of GH secretion or to the release of biogenic amines, or whether it is secondary to 'nonspecific' stress effect produced either by withdrawal or by ethanol in the rat mother.

In conclusion, these data indicate that exposure to ethanol, either chronically or daily at moderate dose during gestation, causes alterations in the regulation of GH secretion in the developing rats. A single dose of ethanol given to pups postnatally also causes marked decline in serum GH levels demonstrating that in the developing rats ethanol can interfere acutely with GH release. It is possible that a suppression of GH release by ethanol may contribute to the 'failure to thrive' growth pattern evident in the fetal alcohol syndrome.

Maternal ethanol exposure has been shown to alter steroidal hormones in the offspring. *Kakihana* et al. [38] have shown that maternal ethanol ingestion alters brain and plasma levels of dihydrosterone and corticosterone in 1- to 2-day-old offspring. However, in a preliminary study, *Taylor* et al. [76] reported that plasma corticosteroid levels are not altered in 1-week-old pups exposed to ethanol continuously. This study also reports that female offspring exposed to ethanol in utero are more sensitive to ethanol as adults. It has been shown that alterations in steroidal hormones in the early neonatal period may cause abnormal neural development in neonatal animals. It appears that further studies are needed to determine the mechanism underlying functional deficits caused by ethanol in the offspring.

Conclusion

From the clinical evidence described in this review, it is clear that intake of alcohol by a pregnant woman causes teratogenic effects in the infant which can range from facial abnormalities, retardation in general growth and development, to functional and neurological deficits.

From the experimental studies, it is also clear that alcohol, given to pregnant animals in doses which are not associated with classical teratogenesis, is still able to cause developmental alterations at the biochemical, neurochemical and hormonal levels. The behavioral alterations seen in the offspring may be due to the changes in neurotransmitters and/or hormones. Also, the retardation in general growth and development may be due to abnormal hormone secretions. Since the current knowledge on the normal processes regulating development of these systems is poorly understood, it is difficult to state the mechanistic action of alcohol on these systems.

Summary

From clinical and experimental studies it is evident that maternal alcohol intake produces deleterious effects on the development of offspring. In infants, these effects can range from lowered birth weight, general retardation of growth and development with functional deficits, to metal retardation with fetal alcohol syndrome. In animals, exposure to alcohol at a level not associated with classical teratological effects can still cause alterations in neural/synaptic development and hormonal secretion. Growth deficiencies and behavioral alterations have also been observed in pups exposed to ethanol in utero. The mechanisms underlying these actions of alcohol are not yet known because the factors that regulate normal growth and development of the central system are still poorly understood.

References

1 Abel, E.L.: Alcohol ingestion in lactating rats: effects on mothers and offspring. 1. Archs int. Pharmacodyn. Thér. *210:* 121–127 (1974).
2 Abel, E.L.: Effects of ethanol on pregnant rats and their offspring. Psychopharmacology, Berlin *57:* 5–11 (1978).
3 Abel, E.L.: Effect of alcohol withdrawal and undernutrition on cannabalism of rat pups. Behav. Neural. Biol. *25:* 411–413 (1979).
4 Abel, E.L.; Kintcheff, B.A.: Effect of prenatal alcohol exposure on growth and development in rats. J. Pharmac. exp. Ther. *207:* 916–921 (1978).
5 Abraham, K.A.: Studies on DNA-dependent RNA polymerase for *Escherichia coli.* I. The mechanism of polyamine-induced stimulation of enzyme activity. Eur. J. Biochem. *5:* 143–146 (1968).
6 Anderson, T.R.; Schanberg, S.M.: Ornithine decarboxylase activity in developing rat brain. J. Neurochem. *19:* 1471–1481 (1972).
7 Anderson, T.R.; Schanberg, S.M.: Effect of thyroxine and cortisol on brain ornithine decarboxylase activity and swimming behaviour in developing rats. Biochem. Pharmac. *24:* 495–501 (1975).

8   Anderson, W.J.; Sides, G.R.: Alcohol induced defects in cerebellar development in the rat; in Galanter, Currents in alcoholism, vol. V, pp. 135–153 (Grune & Stratton, New York 1978).
9   Auroux, M.; Dehaupers, M.: Influence de la nutrition de la mère sur le développement tardif du système nerveux central de la progéniture. C.r. Séanc. Soc. Biol. *164:* 1432–1436 (1970).
10 Ballard, P.L.; Williams-Ashman, H.G.: Isolation and properties of a testicular ribonucleic acid polymerase. J. biol. Chem. *241:* 1602–1615 (1966).
11 Bark, N.: Fertility and offspring of alcoholic women: an unsuccessful search for fetal alcohol syndrome, Br. J. Addict. Alcohol. *74:* 43–49 (1979).
12 Bauer-Moffett, C.; Altman, J.: Ethanol-induced reductions in cerebellar growth of infant rats. Expl Neurol. *48:* 378–383 (1975).
13 Bauer-Moffett, C.; Altman, J.: The effect of ethanol chronically administered to pre-weanling rats on cerebellar development: a morphological study. Brain Res. *119:* 249–268 (1977).
14 Bond, N.W.; Digiusto, E.L.: Effects of prenatal alcohol consumption on open-field behaviour and alcohol preference in rats. Psychopharmacologia, Berlin *46:* 163–165 (1976).
15 Bond, N.W.; Digiusto, E.L.: Effects of prenatal alcohol consumption on shock avoidance learning in rats. Psychol. Rep. *41:* 1269–1270 (1977).
16 Buckalew, L.W.: Effect of maternal alcohol consumption during nursing on offspring activity. Res. Commun. Psychol. Psychiat. Behav. *3:* 353–358 (1978).
17 Butler, S.R.; Schanberg, S.M.: Effect of maternal morphine administration on neonatal rat brain ornithine decarboxylase. Biochem. Pharmac. *24:* 1915–1918 (1975).
18 Branchey, L.; Friedhoff, A.J.: The influence of ethanol administered to pregnant rats on tyrosine hydroxylase activity of their offspring. Psychopharmacologia, Berlin *32:* 151–156 (1975).
19 Chernoff, C.F.: A mouse model of the fetal alcohol syndrome. Teratology *11:* 14A (1975).
20 Chernoff, C.F.: The fetal alcohol syndrome in mice. An animal model. Teratology *15:* 223–230 (1977).
21 Clarren, S.K.; Smith, D.W.: Fetal alcohol syndrome. New Engl. J. Med. *298:* 1063–1067 (1978).
22 Collu, R.; Franschini, F.; Visconti, P.; Martini, L.: Adrenergic and serotoninergic control of growth hormone secretion in adult male rats. Endocrinology *90:* 1231–1237 (1972).
23 Demers, M.; Kirouac, G.: Prenatal effects of ethanol on the behavioural development of the rat. Physiol. Psychol. *6:* 517–520 (1978).
24 Dexter, J.D.; Tumbleson, M.E.; Decker, J.D.; Middleton, C.C.: Fetal alcohol syndrome in Sinclair (S-1) miniature swine. Alcohol. clin. exp. Res. *4:* 146–151 (1980).
25 Diaz, J.; Samson, H.H.: Impaired brain growth in neonatal rat exposed to ethanol. Science, N.Y. *208:* 751–753 (1980).
26 Druse, M.J.; Hofteig, J.H.: The effect of chronic maternal alcohol consumption on the development of central nervous system myelin subfractions in rat offspring. Drug Alcohol Depend. *2:* 421–429 (1977).
27 Duncan, R.J.S.; Woodhouse, B.: The lack of effect in liver alcohol dehydrogenase in mice of early exposure to ethanol Biochem. Pharmac. *27:* 2755–2756 (1978).

28   Ellis, F.W.; Pick, J.R.: An animal model of the fetal alcohol syndrome in beagles. Alcohol. clin. exp. Res. *4:* 123–124 (1980).
29   Ho, B.T.; Fritchie, G.E.; Idanpoan-Heikkila, J.E.; Moissac, W.M.: Placental transfer and tissue distribution of ethanol-1 ^{14}C. Q. Jl Stud. Alcohol *33:* 485–493 (1972).
30   Hofteig, J.H.; Druse, M.J.: Central nervous system myelination in rats exposed to ethanol in utero. Drug Alcohol Depend. *3:* 429–434 (1978).
31   Hanson, J.W.; Jones, K.L.; Smith, D.W.: Fetal alcohol syndrome: experience with 41 patients. J. Am. med. Ass. *235:* 1458–1460 (1960).
32   Henderson, G.I.; Schenker, S.: The effect of maternal alcohol consumption on the viability and visceral development of the newborn rat. Res. Commun. chem. Pathol. Pharmacol. *16:* 15–32 (1977).
33   Henderson, G.I.; Hotumpa, A.M.; Rothschild, M.A.; Schenker, S.: Effect of ethanol and ethanol-induced hypothermia on protein synthesis in pregnant and fetal rats. Alcohol. clin. exp. Res. *4:* 165–177 (1980).
34   Jacobson, S.; Rich, J.; Tousky, N.J.: Delayed myelination and lamination in the cerebral cortex of the albino rat as a result of the fetal alcohol syndrome; in Galanter, Currents in alcoholism, vol. V, pp. 123–133 (Grune & Stratton, New York 1978).
35   Jones, K.L.; Smith, D.W.: Recognition of the fetal alcohol syndrome in early infancy. Lancet *ii:* 999–1001 (1973).
36   Jones, K.L.; Smith, D.W.; Streissguth, A.P.; Myrianthopoulos, M.D.: Outcome in offspring of chronic alcoholic women. Lancet *i:* 1076–1078 (1974).
37   Jones, K.L.; Smith, D.W.: The fetal alcohol syndrome. Teratology *12:* 1–10 (1975).
38   Kakihana, R.; Butte, J.C.; Moore, J.A.: Endocrine effects of maternal alcoholization: plasma and brain testosterone, dihydrotestosterone, estradiol and corticosterone. Alcohol clin. exp. Res. *4:* 57–61 (1980).
39   Kaminski, M.; Rumeau-Rouquette, C., Schwartz, D.: Res. Epidemiol Med. Soc. Sante Publ. *24:* 27–40 (1976); Little R.E.; Schinzel, A.: Alcohol. clin. exp. Res. *2:* 155–159 (1978).
40   Khawaja, J.A.; Wallgren, H.; Usmi, H.; Hilska, P.: Neuronal and liver protein synthesis in the developing offsprings following treatment of pregnant rats with ethanol or 1,3-butanediol. Res. Commun. chem. Pathol. Pharmacol. *22:* 573–580 (1978).
41   Kronick, J.B.: Teratogenic effects of ethyl alcohol administered to pregnant mice. Am. J. Obstet. Gynec. *124:* 676–680 (1976).
42   Krsiak, M.; Ellis J.; Poschlova, N.; Masek, K.: Increased aggresiveness and lower brain serotonin levels in offspring of mice given alcohol during gestation. Q. Jl Stud. Alcohol *38:* 1696–1704 (1977).
43   Lau, C.; Thadani, P.V.; Schanberg, S.M.; Slotkin, T.A.: Effects of maternal ethanol ingestion on development of adrenal catecholamines and dopamine-β-hydroxylase in the offspring. Neuropharmacology *15:* 505–507 (1976).
44   Lemoine, P.; Harrousseau, H.; Borteyru, J.P.; Menuet, J.S.: Les enfants de parents alcooliques: anomalies observées à propos de 127 cas. Ouest méd. *25:* 476–479 (1968).
45   Little, R.E.: Moderate alcohol use during pregnancy and decreased infant birth weight. Am. J. publ. Hlth *67:* 1154–1160 (1977).
46   Little, R.E.: Moderate alcohol use during pregnancy and decreased infant birth weight. Obstetl gynec. Surv. *33:* 710–712 (1978).

47 Martin, J.C.; Martin, D.C.; Sigman, G.; Radow, B.: Offspring survival, development and operant performance following maternal ethanol consumption. Dev. Psychobiol. *10:* 435–446 (1977).
48 Muller, E.E.; Dal Pre, P.; Pecile, A.: Influence of brain neurohumors injected into the lateral ventricle of the rat on growth hormone release. Endocrinology *83:* 893–896 (1968).
49 National Institute of Alcohol Abuse and Alcoholism: Critical review of the fetal alcohol syndrome (1977).
50 Ng, K.Y.; Chase, T.N.; Colburn, R.W.; Kopin, I.J.: *L*-Dopa induced release of cerebral monoamines. Science, N.Y. *70:* 76–77 (1970).
51 Nice, L.B.: Comparative studies of the effects of alcohol, nicotine, tobacco smoke and caffeine on white mice. J. exp. Zool. *12:* 133–152 (1921).
52 Ouellette, E.M.; Rosett, H.L.; Rosman, P.; Weiner, L.: Adverse effects on offspring of maternal alcohol abuse during pregnancy. New Engl. J. Med. *297:* 528–530 (1977).
53 Pearce, L.A.; Schanberg, S.M.: Histamine and spermidine content in brain during development. Science, N.Y. *166:* 1301–1303 (1969).
54 Phillips, D.S.; Stainbrook, G.L.: Effect of early alcohol exposure upon adult ability and taste preferences. Physiol. Psychol. *4:* 473–475 (1976).
55 Randall, C.L.: Teratogenic effect of in utero ethanol exposure; in Blum, Alcohol and opiates, pp. 91–107 (Academic Press, New York 1977).
56 Randall, C.L.; Taylor, W.J.; Walker, D.W.: Ethanol-induced malformations in mice. Alcohol. clin. exp. Res. *1:* 219–224 (1977).
57 Raina, A.; Janne, J.: Polyamines and the accumulation of RNA in mammalian systems. Fed. Proc. *29:* 1568–1574 (1970).
58 Raina, A.; Janne, J.: Physiology of the natural polyamines, putrescine spermidine and spermine. Med. Biol. *53:* 121–147 (1975).
59 Rawat, A.K.: Ribosomal protein synthesis in the fetal and neonatal rat brain as influenced by maternal ethanol consumption. Res. Commun. chem. Pathol. Pharmacol. *12:* 723–732 (1975).
60 Rawat, A.K.: Effect of maternal ethanol consumption of foetal and neonatal rat hepatic protein synthesis. Biochem. J. *160:* 653–661 (1976).
61 Rawat, A.K.: Effect of maternal ethanol consumption on hepatic *lipid* biosynthesis in foetal and neonatal rats. Biochem. J. *174:* 213–219 (1978).
62 Riley, E.P.; Lochry, E.A.; Shapiro, N.R.: Lack of response inhibition in rats prenatally exposed to alcohol. Psychopharmacology *62:* 47–52 (1979).
63 Riley, E.P.; Lochry, E.A.; Shapiro, N.R.; Baldwin, J.: Response perserveration in rats exposed to alcohol prenatally. Pharmacol. Biochem. Behav. *10:* 255–259 (1979).
64 Russell, D.H.; Snyder, S.H.: Amine synthesis in rapidly growing tissues: ornithine decarboxylase activity in regenerating rat liver, chick embryo and various tumors. Proc. natn. Acad. Sci. USA *60:* 1420–1427 (1968).
65 Sandor, S.: The influence of ethyl alcohol on the developing chick embryo. Revue roum. Embryol *5:* 167–171 (1968).
66 Sandor, S.; Amels, D.: The action of ethanol on the prenatal development of albino rats. Revue roum. Embryol. *8:* 105–118 (1971).
67 Shaw, G.G.: The polyamines in the central nervous system. Biochem. Pharmac. *28:* 1–6 (1979).

68 Sjöblom, M.; Oisund, J.F.; Morland, J.: Development of alcohol dehydrogenase and aldehyde dehydrogenase in the offspring of female rats chronically treated with ethanol. Acta pharmac. tox. *44:* 128–131 (1979).
69 Smythe, G.S.; Branstater, J.; Lazarus, L.: Serotoninergic control of rat growth hormone secretion. Neuroendocrinology *17:* 245–257 (1975).
70 Smythe, G.A.; Lazarus, L.: Growth hormone regulation by melatonin and serotonin. Nature, Lond. *244:* 230–231 (1973).
71 Streissguth, A.P.; Landesman-Dwyer, S.; Martin, J.C.; Smith, D.W.: Teratogenic effects of alcohol in humans and laboratory animals. Science, N.Y. *209:* 353–361 (1980).
72 Stuart, M.; Lazarus, L.; Smythe, G.A.; Moore, S.; Sara, V.: Biogenic amine control of growth hormone secretion in the fetal and neonatal rat. Neuroendocrinology *22:* 337–342 (1977).
73 Sullivan, W.C.: A note on the influence of maternal inebriety on the offspring. J. ment Sci. *45:* 489–493 (1899).
74 Swanberg, K.M.; Wilson, J.R.; Kalisker, A.: Developmental and genotypic effects on pituitary-adrenal function and alcohol tolerance in mice. Dev. Psychobiol. *12:* 201–210 (1979).
75 Tabor, H.; Tabor, C.W.: Spermidine, spermine and related amines. Pharmac. Rev. *16:* 245–300 (1964).
76 Taylor, A.N.; Branch, B.J.; Lui, S.; Kokka, N.: Fetal exposure to alcohol enhances pituitary-adrenal and hypothermic responses to alcohol in adult rats. Alcohol. clin. exp. Res. *4:* 231 (1980).
77 Thadani, P.V.: Effect of a single dose of ethanol administered to pregnant rats on developing brain and heart ornithine decarboxylase activity. Fed. Proc. *39:* 542 (1980).
78 Thadani, P.V.: Response of heart and brain ornithine decarboxylase activity in the developing rats exposed to ethanol either during mid or late gestation period. Alcohol clin. exp. Res. *4:* 231 (1980).
79 Thadani, P.V.; Kulig, B.M.; Brown, F.C.; Beard, J.D.: Acute and chronic ethanol-induced alterations in brain norepinephrine metabolites in the rat. Biochem. Pharmac. *25:* 93–95 (1976).
80 Thadani, P.V.; Lau, C.; Slotkin, T.A.; Schanberg, S.M.: Effects of maternal ethanol ingestion on amine uptake into synaptosomes of fetal and neonatal brain. J. Pharmac. exp. Ther. *200:* 292–297 (1977).
81 Thadani, P.V.; Lau, C.; Slotkin, T.A.; Schanberg, S.M.: Effect of maternal ethanol ingestion on neonatal rat brain and heart ornithine decarboxylase. Biochem. Pharmac. *26:* 523–527 (1977).
82 Thadani, P.V.; Slotkin, T.A.; Schanberg, S.M.: Effects of late prenatal or early postnatal ethanol exposure on ornithine decarboxylase activity in brain and heart of developing rats. Neuropharmacology *16:* 289–293 (1977).
83 Thadani, P.V.; Schanberg, S.M.: Effect of maternal ethanol ingestion on serum growth hormone in the developing rat. Neuropharmacology *18:* 821–826 (1979).
84 Thadani, P.V.; Truitt, E.B.: Effect of acute ethanol or acetaldehyde administration on the uptake, release, metabolism and turnover rate of norepinephrine in rat brain. Biochem. Pharmac. *26:* 1147–1150 (1977).

85 Tze, W.J.; Friesen, H.G.; MacLeod, P.M.: Growth hormone response in fetal alcohol syndrome. Archs. Dis. Childh. *51:* 703–706 (1976).
86 Tze, W.J.; Lee, M.: Adverse effects of maternal alcohol consumption on pregnancy and foetal growth in rats. Nature, Lond. *257:* 479–480 (1975).
87 Vincent, N.M.: The effects of prenatal alcoholism upon motivation, emotionality and learning. Am. Psychol. *13:* 401–405 (1958).
88 Waltman, R.; Iniquez, E.S.: Placental transfer of ethanol and its elimination at term. Obstet. Gynec. *40:* 185–189 (1972).

Dr. P.V. Thadani, Ph.D., Veterans Administration Medical Center,
Washington, DC 20422 (USA)

Neuroendocrine Effects of Fetal Alcohol Exposure[1]

Anna Newman Taylor, Berrilyn J. Branch, Norio Kokka

Brentwood VA Medical Center, and Department of Anatomy and Laboratory of Neuroendocrinology, Brain Research Institute, University of California, Los Angeles, Calif., USA

Introduction

On the basis of ample clinical evidence it is now generally accepted that excessive maternal consumption of alcohol during pregnancy can produce various adverse affects on the developing fetus and neonate [29]. Exposure of the human fetus to alcohol may result in growth deficiencies, genetic abnormalities and/or signs of central nervous system (CNS) dysfunction. Indeed, a distinct dysmorphic condition, recognized as the fetal alcohol syndrome (FAS), is characterized by a typical cluster of craniofacial abnormalities, failure to thrive and mental retardation [8]. The teratogenic effects of alcohol have also been demonstrated in laboratory animals [26, 29].

Thus, while it is clear that fetal exposure to alcohol affects the developing organism, the mechanisms whereby alcohol produces teratogenic effects remain essentially unknown. Nor have we yet assessed the full extent of the physiological involvement and the duration of the effects, i.e. whether the fetal alcohol-induced alterations persist into adulthood. Alcohol or its metabolites may act directly on the developing fetus and depending upon dose and time of exposure, differential ontogenetic effects may occur. On the other hand, maternal alcohol consumption induces a complex of interactive factors such as undernutrition and poor maternal behavior, which also affect ontogenetic processes in the fetus and neonate. Alcohol may also exert effects on maternal and/or fetal

---

[1] The authors' research is supported by the Veterans Administration Medical Research Service.

physiological systems, such as the endocrine system, whose perturbation may in turn influence ontogenetic processes.

The purpose of this chapter is to address some of the issues raised above concerning the range of effects and mechanisms of action of fetal alcohol exposure. First, we will review recent results, primarily from other laboratories (our experiments are still in progress), on the effects of fetal alcohol exposure on neuroendocrine functions. Then, we will describe experiments from our laboratory which demonstrate persistent effects on pituitary-adrenal and body temperature responses to a challenge dose of alcohol in adult rats exposed to alcohol in utero. Finally, based on our own work and that of others, we will develop the hypothesis that alcohol-induced activation of the hypothalamo-pituitary-adrenal (HPA) axis during gestation may contribute to the adverse effects of fetal alcohol exposure.

Early Neuroendocrine Effects of Fetal Alcohol Exposure

Various studies have demonstrated that alcohol causes abnormal endocrine function in adult humans and animals [38]. Alcohol has been shown to act at all levels of the neuroendocrine axis, i.e. hypothalamus, pituitary, endocrine glands, target organs. Does fetal exposure to alcohol also affect neuroendocrine function in neonates? Only a few reports have appeared thus far to answer this question.

Pituitary-Adrenal Function

Rats exposed to ethanol during the last week of gestation were found to have elevated brain levels of corticosterone at birth [19]. Plasma corticosterone levels also tended to be elevated in the newborns. Since ethanol is known to activate the HPA axis of adult animals and humans [38], the source of the elevated corticosteroids in the neonate could well be maternal. However, the ethanol-treated pups tended to have enlarged adrenal glands indicating that the fetal HPA axis appeared to be activated as well.

A 4-month-old infant has been described who developed a pseudo-Cushing's syndrome as a result of exposure to alcohol in breast milk [4]. While this is the first such report in infants, adult alcoholics are known to develop physical characteristics similar to those of Cushing's syndrome [38]. We are in the process of extending our preliminary observations of alterations in basal corticosteroid secretory patterns during the preweaning period in fetally exposed rats [9]. Implications of elevated corticosteroid levels in the neonate for somatic and cerebral developmental processes are discussed below.

Growth Hormone (GH) Secretion

In two case studies of FAS children of varying ages (0.4–15 years) GH responses to insulin-induced hypoglycemia and arginine infusion were normal [27, 35]. Although these studies appear to indicate normal GH function in FAS offspring, a recent animal study on basal GH levels suggests impaired function. Following an initial increase at birth, serum GH levels fall and remain reduced in the preweaning period in rats exposed to alcohol during the last week of gestation [34]. Acute or chronic postnatal preweaning exposure similarly reduced GH levels. These preweaning reductions in basal secretion of GH may contribute to the growth retardation observed in the FAS, although a contributory role of nutritional factors cannot be excluded.

Pituitary-Thyroid Function

One study has demonstrated a reduction in serum thyroxine levels on postnatal day 11 in rats exposed to ethanol in utero [20]. Although the levels of this hormone were normal on day 14, it is important to consider the possibility that the hypothalamo-pituitary-thyroid axis may be affected by prenatal alcohol exposure. Impaired function of this neuroendocrine axis may also affect the metabolic status of the developing neonate. Furthermore, this study demonstrated that fetal alcohol exposure, like hypothyroidism [3], delays cerebral development.

Reproductive Function

Prenatal exposure of mice to alcohol during days 5–11 of gestation has been found to delay sexual maturation as measured by the time of vaginal opening [5]. These investigators indicate that the delay probably reflects a combination of food restriction and an action of alcohol per se during gestation. In one study we also observed a significant delay in vaginal opening in rats exposed to ethanol in utero (days 8–21 of gestation) when compared with normally fed rats, but not when compared with isocaloric pair-fed controls [unpublished observations]. Thus our results also support a contribution of prenatal nutritional status to this effect. In another study, however, adult female mice which had been exposed to alcohol in utero and until 14 days of age displayed normal reproductive ability and fertility [15]. Additional indices of sexual maturation will have to be examined before a definitive statement can be made about the effects of fetal alcohol exposure on reproductive function in the female.

Brain and plasma estradiol levels were found to be unaffected in newborn male and female pups exposed to alcohol during the 3 RD week of gestation [19]. However, in the same study brain levels of dihydrotestosterone in male pups at birth were reduced by the alcohol exposure. Testosterone levels in brain and plasma were unaltered, but there was a tendency toward reduced testicular weight and increased adrenal weight of neonatal male rats. Another recent study examined sexual differentiation in male rats exposed to alcohol during the latter two thirds of pregnancy and/or the 1st week postnatally [7]. Prenatal treatment masculinized the anogenital distance of both sexes at birth, while puberty occurred significantly earlier in postnatally treated males than in prenatally treated animals or controls. In the male offspring in adulthood alcohol reduced penile reflexes, but was without effect on plasma testosterone levels and weight of sex accessory glands. Although the effects on mating behavior were not significant, there was a tendency for fewer pre- or postnatally alcohol-treated animals to show mating behavior and postnatally treated males showed less sexual behavior. The results of these studies indicate that alcohol treatment during pregnancy exerts effects on sexual differentiation of the offspring. The findings of *Kakihana* et al. [19] on brain dihydrotestosterone levels are significant in view of the important role of testosterone in differentiation of the male brain during the perinatal period [16].

Adult Neuroendocrine and Body Temperature Effects of Fetal Alcohol Exposure

The purpose of our study of some physiological effects of fetal alcohol exposure was twofold. First, we sought to determine whether perinatal exposure to alcohol would produce long-lasting neuroendocrine effects in the organism which persist to adulthood. Although clinical studies have thus far generally not extended beyond teenagers, experiments in animals have demonstrated persistent neurobehavioral effects of fetal alcohol exposure [30]. Thus, it was of interest to examine neuroendocrine function in adultanimals exposed to alcohol in utero. Secondly, in these studies we focused on alcohol-induced neuroendocrine and body temperature responses in order to determine whether perinatal exposure to alcohol produces long-lasting tolerance to this drug as we had previously observed with morphine in adult rats neonatally exposed to the opioid [42].

The experiments utilized adult Sprague-Dawley rats, 52–90 days of age, which had been exposed in utero (weeks 2 and 3 of gestation) or postnatally (week 1) to alcohol. Their mothers were fed a liquid diet (Bio-Serv, Frenchtown, N.J.) which contained 5.0% w/v ethanol, ad libitum. Control dams were pair-fed an isocaloric liquid diet with maltose-dextrin replacing the ethanol. At birth the ethanol- and pair-fed derived pups were randomized and cross-fostered in litters of 5 males and 5 females each to normal dams (fed lab chow and water ad libitum). The offspring of the normal dams were cross-fostered to the ethanol and pair-fed dams (After the first week of lactation these dams were replaced on the lab chow/water diets.) Cross-fostering was instituted to permit dissociation of the contributions of pre- and postnatal exposure. This procedure yielded four groups of experimental animals: ethanol in utero – normal postnatal (EN), pair-fed in utero – normal postnatal (PN), normal in utero – ethanol postnatal week 1 (NE), and normal in utero – pair-fed postnatal week 1 (NP). On the day of birth the only observable effect of in utero alcohol exposure was that such pups weighed significantly less than pair-fed or normal, untreated pups, but these differences disappeared by day 14; at the time of the experiments reported here there were no differences in body weight between the groups.

Pituitary-Adrenal Response to Alcohol

At 90 days of age, female rats were tested for their pituitary-adrenal response to an i.p. injection of ethanol (1.0 g/kg, 20% w/v saline) or an equal volume of the saline vehicle. Blood samples drawn at 60 min after the injection were analyzed fluorometrically for corticosterone content. The results reported in table I for rats exposed in utero to ethanol or to the pair-fed control diet demonstrate this effect. In both prenatal diet groups (EN and PN) ethanol significantly elevated corticosterone titers above those of the saline controls at 60 min postinjection. Furthermore, it can be seen from table I that the corticosteroid response to ethanol in fetally exposed EN rats was significantly greater than in the pair-fed controls (PN). At this time, we do not know whether this hyperresponsiveness can be generalized to other stressors or whether the enhanced pituitary-adrenal response of rats fetally exposed to alcohol is specific to ethanol.

Table I. Pituitary-adrenal response to ethanol of female rats exposed to ethanol in utero or during postnatal week 1

Groups	Plasma corticosterone, μg/100 ml, 60 min after			
(in utero – postnatal)	saline	animals n	ethanol 1,0 g/kg, i.p.	animals n
Ethanol – Normal	19.21 ± 2.23^a	8	52.04 ± 7.35^c	7
Pair-fed – Normal	15.26 ± 3.05^b	8	28.04 ± 4.06	7
Normal – Ethanol	17.53 ± 2.61	6	24.59 ± 3.72	7
Normal – Pair-fed	17.07 ± 2.82	6	19.51 ± 3.67	7

Results are means ± SEM.
[a] $p < 0.01$, saline versus ethanol response (by Student's t test).
[b] $p < 0.05$, saline versus ethanol response.
[c] $p < 0.02$, versus ethanol response of pair-fed – normal group.

It is interesting to note that both postnatal treatments rendered the rats less responsive to ethanol, at least at 60 min postinjection. Since the two postnatally treated NE and NP groups did not differ in their corticosteroid responses to ethanol, postnatal factors such as malnutrition and poor maternal behavior, rather than alcohol per se, must have contributed to their response patterns. On the other hand, with in utero exposure, it appears that alcohol per se renders rats hyperresponsive to the pituitary-adrenal activating effects of alcohol as adults.

Alcohol-Induced Hypothermia

Rectal temperatures were measured with a telethermometer at 30-min intervals following i.p. injection of ethanol (1.5 g/kg, 20% w/v saline). Figure 1 shows the temperature response of female rats from the in utero and postnatal treatment group tested at 52 days of age. In utero ethanol-exposed EN rats showed a significantly greater fall in body temperature than pair-fed controls (PN) at 60, 90, 120, 150 and 180 min following ethanol injection. Analysis of variance of the total temperature responses calculated for the entire 180-min period (fig. 1, right-hand panel) of the four treatment groups was significant at the 0.01 level (F 3,31 = 6.33). The total hypothermic response is significantly greater in the prenatally exposed EN group than in the pair-fed (PN) controls (fig. 1, right-hand panel). There was no difference between hypothermic responses in either postnatally treated group (NE and NP) and the re-

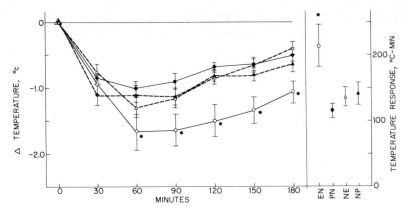

Fig. 1. Temperature responses to ethanol in 52-day-old female rats exposed to 5.0% w/v ethanol in utero (gestation weeks 2 and 3) or during week 1 postnatally. Mean individual responses (Δ °C ± SEM) to ethanol (1.5 g/kg, 20% w/v, i.p.) or saline at 30-min intervals for 180 min postinjection are shown for the in utero exposed (EN and PN) and postnatally treated (NE and NP) groups. Responses of normal untreated rats are not shown, since they did not differ from PN, NE and NP curves. Preinjection body temperatures (0 min) averaged 37.4 °C and did not differ significantly between groups. * = Times when the hypothermic response of the EN group is significantly greater ($p < 0.01$) than that of the PN control group (by Student's t test). In the right-hand panel, the mean ± SEM total hypothermic response (°C-min), calculated for the entire 180-min period (area under the curves for each animal in the left-hand panel), is shown for the four treatment groups. * = The EN response is significantly greater ($p < 0.01$) than the PN response. ○ = EN (n = 10); ● = PN (n = 9); △ = NE (n = 8); ▲ = NP (n = 8).

sponses of these groups were similar to the prenatal pair-fed control (PN) group. Although not shown in figure 1, the alcohol-induced hypothermic responses of normal (untreated) rats do not differ from those of the prenatal pair-fed (PN) group. We have recently demonstrated enhanced hypothermic responses to alcohol in fetal alcohol-exposed male rats at 90 days of age [unpublished observations].

In sum, these results demonstrate that fetal, but not postnatal exposure to alcohol renders adult rats more responsive to the pituitary-adrenal activating and hypothermic effects of a challenge dose of alcohol. These two biologic responses are mediated by ethanol's actions on the thermoregulatory and corticotropin-releasing factor regions of the hypothalamus [12, 14]. Thus, it is conceivable that the sensitivity of the hypothalamic sites to ethanol has in some way been altered during development. Indeed, perinatal ethanol exposure is known to exert effects on brain protein formation and the development of central neurotransmitter systems which also mediate neuroendocrine functions [30]. On the other hand, peripheral systemic or metabolic effects of fetal alcohol exposure cannot be excluded. These results also demonstrate that fetal exposure to alcohol produces persistent effects on the organism's physiologic responses to alcohol, but, contrary to our expectation, the long-lasting effect is enhanced rather than reduced responsivity to the drug. While it has been reported that perinatally exposed rats demonstrate an increased preference for alcohol as adults [6, 25], to our knowledge this is the first report [31] of persistent effects on biologic responses to alcohol in adults (of any species) exposed to the drug in utero.

Possible Role of Pituitary-Adrenal Activation in Fetal Alcohol Effects

Both the development of the young rat and its adult behavior can be influenced by exposing infant rats to various hormones or stressful treatments [1, 11, 23]. The common denominator of many of the various effective perinatal treatments appears to involve alteration of the pituitary-adrenal system. Based on the demonstrated organizational influences of the gonadal hormones on the CNS during critical periods of development [16], it has been suggested that the pituitary-adrenal hormones may exert similar organizing effects on the developing CNS [23].

Treatment of newborn rodents with pharmacological doses of corticosteroids has been shown to inhibit somatic and brain growth and cellular differentiation, proliferation and myelination in the CNS of the neonate [17, 24, 28, 39]. Neonatal corticosteroid exposure affects brain levels of the biogenic amines [36] and their ontogenetic patterns [22], and CNS development of enzymes for serotonin and tryptophan metabolism [41] and brain activity of ornithine decarboxylase [2]. Development of the circadian pituitary-adrenal rhythm is delayed and adult amplitudes of the rhythm are reduced [10, 21, 32, 33, 37]. There are also long-term alterations in behavior [18, 28] and stress responsiveness [13, 37]. Thus, it is apparent that perinatal glucocorticoid therapy during critical periods of CNS development can exert immediate and irreversible effects on brain cell division and differentiation and neurochemical processes which result in long-term physiological and behavioral consequences [40].

It is significant that many, if not all, of the systems affected by corticosteroids are similarly affected by fetal alcohol exposure [30]. Indeed, evidence was presented above for elevated brain levels of corticosteroids and possible adrenal stimulation following fetal alcohol exposure [19]. Thus, the possibility that alcohol-induced activation of the HPA axis may contribute to the actions of alcohol on the developing organism deserves serious attention. We have begun to test this hypothesis. We found that rats treated with ACTH neonatally display exaggerated hypothermic responses to a challenge dose of alcohol as adults [unpublished observations], as we discussed above for fetal alcohol-exposed rats. The apparent similarity between the effects of these two perinatal treatments suggests that increased ACTH secretion may be common to both.

Concluding Remarks

The neuroendocrine effects of fetal exposure to alcohol have only recently begun to be investigated. The effects demonstrated thus far in newborn rodents include elevated brain corticoid levels and reduced plasma thyroxine and growth hormone levels. While we singled out for discussion the effects of elevated corticosteroid levels on somatic and cerebral developmental processes because of our investigations in this area, we do not intend to imply that reductions in thyroxine and GH levels cannot also exert profound influences on ontogeny. Sexual differentiation is also found to be affected. On the basis of such information concerning effects on neuroendocrine systems appropriate hormonal therapy could be designed to treat the sequelae of fetal alcohol exposure.

Our demonstration of long-lasting effects on pituitary-adrenal and body temperature responses to alcohol is of interest because it indicates that biologic effects induced during development by fetal alcohol exposure can persist into adulthood. Furthermore, our finding of a persistent effect on the organism's sensitivity to alcohol may be significant for predicting risks for alcoholism in fetally exposed adults. Additional experiments in which both basal and stimulated hormonal secretory patterns are investigated in adult animals and humans are required to substantiate the long-term physiological consequences of fetal alcohol exposure.

The need for more information on the neuroendocrine effects exerted by fetal alcohol exposure is readily apparent. Such information would be useful not only for assessment of the full extent of the effects of fetal alcohol exposure on the developing organism, but also for an understanding of the mechanisms of alcohol's actions.

Summary

While the actions of alcohol on endocrine function in adults are well-documented, the effects of fetal exposure to alcohol on neuroendocrine function in neonates and adults are only beginning to be investigated. Recent reports are reviewed which demonstrate effects of fetal alcohol exposure on pituitary-adrenal function, GH secretion, thyroxine levels and sexual differentiation in newborn rodents. Our studies of the long-term effects of fetal and early postnatal exposure to alcohol on pituitary-adrenal and body temperature responses to a challenge dose of ethanol in adult rats are described. Both responses are enhanced in prenatally, but not in postnatally exposed rats, indicating that the effects of fetal alcohol exposure on physiological systems, such as the endocrine and thermoregulatory systems, persist to adulthood. Based on apparent similarities in the somatic and cerebral deficits which occur following fetal alcohol exposure and neonatal corticosteroid treatment, a hypothesis is developed for the role of alcohol-induced activation of the HPA axis during gestation in the adverse effects of fetal alcohol exposure.

References

1   Ader, R.; Grota, L.J.: Adrenocortical mediation of the effects of early life experiences; in Zimmermann, Gispen, Marks, DeWied, Drug effects on neuroendocrine regulation. Prog. Brain Res. *39:* 395–405 (1973).
2   Anderson, T.R.; Schanberg, S.M.: Effect of thyroxine and cortisol on brain ornithine decarboxylase activity and swimming behavior in developing rat. Biochem. Pharmac. *24:* 495–501 (1975).
3   Balazs, R.; Patel, A.J.; Hajos, F.: Factors affecting the biochemical maturation of the brain: effects of hormones during early life. Psychoneuroendocrinology *1:* 25–36 (1975).
4   Binkiewicz, A.; Robinson, M.J.; Senior, B.: Pseudo-Cushing syndrome caused by alcohol in breast milk. J. Pediat. *93:* 965–967 (1978).
5   Bogan, W.O.; Randall, C.L.; Dodds, H.M.: Delayed sexual maturation in female C57BL/6J mice prenatally exposed to alcohol. Res. Commun. chem. Pathol. Pharmacol. *23:* 117–125 (1979).
6   Bond, N.W.; DiGuisto, E.L.: Effects of prental alcohol consumption on open-field behavior and alcohol preference in rats. Psychopharmacology *46:* 163–165 (1976).
7   Chen, J.J.; Smith, E.R.: Effects of perinatal alcohol on sexual differentiation and open-field behavior in rats. Horm. Behav. *13:* 219–231 (1979).
8   Clarren, S.K.; Smith, D.W.: The fetal alcohol syndrome. New Engl. J. Med. *298:* 1063–1067 (1978).
9   Cooley-Matthews, B.; Taylor, A.N.: Exposure to ethanol in utero may delay maturation of hypothalamo-pituitary-adrenal function in the rat. Abstr. Soc. Neurosci. *4:* 110 (1978).
10 Cost M.G.; Mann, D.R.: Neonatal corticoid administration: retardation of adrenal rhythmicity and desynchronization of puberty. Life Sci. *19:* 1929–1936 (1976).

11   Denenberg, V.H.; Zarrow, M.S.: Effects of handling in infancy upon adult behavior and adrenocortical activity: suggestions for a neuroendocrine mechanism; in Walcher, Peters, Early childhood: the development of self-regulatory mechanisms, pp. 39–64 (Academic, New York 1971).
12   Ellis, F.W.: Effect of ethanol on plasma corticosterone levels. J. Pharmac. exp. Ther. 153: 121–127 (1966).
13   Erskine, M.S.; Geller, E.; Yuwiler, A.: Effects of neonatal hydrocortisone treatment on pituitary and adrenocortical response to stress in young rats. Neuroendocrinology 29: 191–199 (1979).
14   Freund, G.: Alcohol withdrawal syndrome in mice. Archs Neurol. 21: 315–320 (1969).
15   Ginsburg, B.E.; Yanali, J.; Sze, P.Y.: A developmental genetic study of the effects of alcohol consumed by parent mice on the behavior and development of their offspring; in Chafetz, Research, treatment and prevention, pp. 183–204 (NIAAA, Rockville, Md. 1975).
16   Gorski, R.A.: Perinatal effects of sex steroids on brain development and function; in Zimmermann, Gipsen, Marks, DeWied, Drug effects on neuroendocrine regulation. Prog. Brain Res. 39: 149–163 (1973).
17   Gumbinas, M.; Oda, M.; Huttenlocher, P.: The effects of corticosteroids on myelination of the developing rat brain. Bio. Neonate 22: 355–366 (1973).
18   Howard, E.; Granoff, D.M.: Increased voluntary running and decreased motor coordination in mice after neonatal corticosterone implantation. Expl. Neurol. 22: 661–673 (1968).
19   Kakihana, R.; Butte, J.C.; Moore, J.A.: Endocrine effects of maternal alcoholization: plasma and brain testosterone, dihydrotestosterone, estradiol, and corticosterone. Alcoholism 4: 57–60 (1980).
20   Kornguth, S.E.; Rutledge, J.J.; Sunderland, E.; Siegel, F.; Carlson, I.; Smollens, J.; Juhl, U.; Young, B.: Impeded cerebellar development and reduced serum thyroxine levels associated with fetal alcohol intoxication. Brain Res. 177: 347–360 (1979).
21   Krieger, D.T.: Circadian corticosteroid periodicity: critical period for abolition by neonatal injection of corticosteroid. Science 178: 1205–1207 (1972).
22   Lengvari, I.; Branch, B.J.; Taylor, A.N.: Effects of perinatal thyroxine and/or corticosterone treatment on the ontogenesis of hypothalamic and mesencephalic norepinephrine and dopamine content. Dev. Neurosci. 3: 59–65 (1980).
23   Levine, S.: The pituitary-adrenal system and the developing brain; in DeWied, Weijnen, Pituitary, adrenal and the brain. Prog. Brain Res. 32: 79–85 (1970).
24   Oda, M.A.; Huttenlocher, P.R.: The effect of corticosteroids on dendritic development in the rat brain. Yale J. Biol. Med. 3: 155–165 (1974).
25   Phillips, D.S.; Stainbrook, G.L.: Effects of early alcohol exposure upon adult learning ability and taste preferences. Physiol. Psychol. 4: 473–475 (1976).
26   Randall, C.L.: Teratogenic effects of in utero ethanol exposure; in Blum, Alcohol and opiates, neurochemical and behavioral mechanisms, pp. 91–107 (Academic, New York 1977).
27   Root, A.W.; Reiter, E.O.; Andriola, M.; Duckett, G.: Hypothalamic-pituitary function in the fetal alcohol syndrome. J. Pediat. 87: 585–588 (1975).
28   Schapiro, S.: Some physiological, biochemical and behavioral consequences of neonatal hormone administration: cortisol and thyroxine. Gen. compar. Endocr. 10: 214–228 (1968).

29 Streissguth, A.P.; Landesman-Dwyer, S.; Martin, J.C.; Smith, D.W.: Teratogenic effects of alcohol in humans and laboratory animals. Science 209: 353–361 (1980).
30 Taylor, A.N.: Neurologic sequelae of fetal alcoholism: a review; in Braude, Genetic, perinatal and developmental effects of abused substances (Raven, New York, in press).
31 Taylor, A.N.; Branch, B.J.; Liu, S.; Kokka, N.: Fetal exposure to alcohol enhances pituitary-adrenal and hypothermic responses to alcohol in adult rats. Alcoholism 4: 231 (1980).
32 Taylor, A.N.; Langvari, I.: Effect of combined perinatal thyroxine and corticosterone treatment on the development of the diurnal pituitary-adrenal rhythm. Neuroendocrinology 24: 74–79 (1977).
33 Taylor, A.N.; Lorenz, R.J.; Turner, B.B.; Ronnekleiv, O.K.; Casady, R.L.; Branch, B.J.: Factors influencing pituitary-adrenal rhythmicity, its ontogeny and circadian variations in stress responsiveness. Psychoneuroendocrinology 1: 291–301 (1976).
34 Thadani, P.V.; Schanberg, S.M.: Effect of maternal ethanol ingestion on serum growth hormone in the developing rat. Neuropharmacology 18: 821–826 (1979).
35 Tze, W.J.; Friesen, H.G.; MacLeod, P.M.: Growth hormone response in fetal alcohol syndrome. Archs Dis. Childh. 51: 703–706 (1976).
36 Ulrich, R.; Yuwiler, A.; Geller, E.: Effects of hydrocortisone on biogenic amine levels in the hypothalamus. Neuroendocrinology 19: 259–268 (1975).
37 Ulrich, R.; Yuwiler, A.; Geller, E.: Neonatal hydrocortisone: effect on the development of the stress response and the diurnal rhythm of corticosterone. Neuroendocrinology 21: 49–57 (1976).
38 Van Thiel, D.H.: Alcohol and its effect on endocrine functioning. Alcoholism 4: 44–49 (1980).
39 Vernadakis, A.; Woodbury, D.M.: Effects of cortisol on maturation of the central nervous system; in Ford, Influence of hormones on the nervous system, pp. 85–97 (Karger, Basel 1971).
40 Weichsel, M.E., Jr.: The therapeutic use of glucocorticoid hormones in the perinatal period: potential neurological hazards. Ann. Neurol. 2: 364–366 (1977).
41 Yuwiler, A.; Simon, M.; Bennett, B.; Plotkin, S.; Wallace, R.; Brammer, C.; Ulrich, R.: Effect of neonatal corticoid treatment on tryptophan and serotonin metabolism. Endocrinol. exp. 12: 21–30 (1978).
42 Zimmermann, E.; Branch, B.; Taylor, A.N.; Young, J.; Pang, C.N.: Long-lasting effects of prepubertal administration of morphine in adult rats; in Zimmermann, George, Narcotics and the hypothalamus, pp. 183–196 (Raven Press, New York 1974).

Anna N. Taylor, PhD, Department of Anatomy, School of Medicine,
University of California, Los Angeles, CA 90024 (USA)

Prenatal Effects of Beverage Alcohol on Fetal Growth[1]

Ernest L. Abel[2]

Research Institute on Alcoholism, Buffalo, N.Y., USA

Introduction

Prenatal growth retardation is one of the most reliable effects of in utero alcohol exposure in both animals and humans [1]. The present study determined whether the congeners present in various beverage alcohols affect fetal growth to a greater extent than exposure to ethanol alone.

Materials and Methods

Timed pregnant Long Evans rats (n=5/group) were individually housed in Plexiglas cages and were intubated twice daily with the equivalent of 3 g/kg of ethanol in the form of beer (Budweiser), wine (Almaden Rosé), whiskey (Hiram Walker), or ethanol (USP). All solutions were diluted to a final ethanol concentration of 8% v/v except for beer to which ethanol was added to bring it up to the desired concentration. Concentrations were determined by gas chromatography. A fifth group of dams was intubated with an isocaloric sucrose solution.

Intubations began on day 1 of pregnancy and continued until day of sacrifice on gestation day 20. All dams were pair-fed and watered to dams in the ethanol-treated group.

On day 20, approximately 1 h after drug treatment, dams were sacrificed by decapitation, and a 25-μl sample of blood was obtained. These samples were deproteinized using $Ba(OH)_2$ and $ZnSO_4$, and 25-μl of propanol were added as an internal standard. Blood

---

[1] This research was supported by New York State Health Planning grant HRC No. 9–014.

[2] I thank Dr. *Hebe B. Greizerstein* for blood alcohol analysis and analysis of congener contents in alcoholic beverages, and *B. Dintcheff* and *R. Bush* for technical assistance.

alcohol levels in these samples were subsequently determined by gas chromatography by the method of *Greizerstein and Smith* [4].

Following removal of blood samples, fetuses were removed, blotted, and individually weighed. The data were subjected to analysis of variance. Fetal weights were subjected to analysis of variance for 'nested' data [6] which takes into account the contribution of maternal factors to fetal weight.

Results

The congener content of the three different beverages is presented in figure 1. As indicated by the figure, each beverage contained levels of various congeners that were higher or lower than those present in the

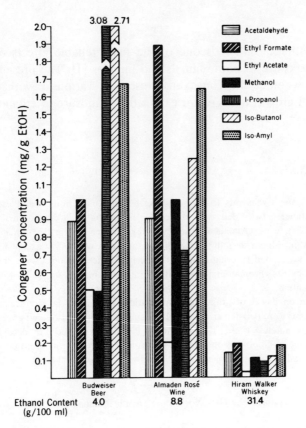

Fig. 1. Congener content of beer (Budweiser), wine (Almaden Rosé), and whiskey (Hiram Walker) [data derived from ref. 5].

Table I. Effects of beer, wine, whiskey, and ethanol on fetal development in rats ($\bar{X} \pm SE$)

Treatment	n	Maternal blood alcohol level mg%	Litter size	Fetal weight g
Beer	5	172 ± 21	8.0 ± 1.6	3.5 ± 0.11
Wine	5	206 ± 21	9.8 ± 1.5	3.2 ± 0.07
Whiskey	5	204 ± 14	12.0 ± 0.6	3.8 ± 0.04
Ethanol	5	204 ± 11	12.5 ± 1.3	4.1 ± 0.04
Control	5		11.6 ± 0.7	4.4 ± 0.07

other beverages. Maternal blood alcohol levels, litter size, and fetal weights are presented in table I. Group differences in blood alcohol levels were not significant, although levels in dams treated with beer were somewhat lower than those in dams exposed to the other solutions. Maternal weights were not significantly different at time of sacrifice. Litter size was also not significantly affected. Group differences in fetal weight were significant (F = 44.5, d.f. = 4.265, p < 0.001). An internal analysis indicated that each of the alcohol-treated fetuses weighed significantly less (p < 0.01) than the control fetuses. Wine- and beer-exposed fetuses, but not whiskey-exposed fetuses, weighed significantly less than ethanol-exposed fetuses (p < 0.01). Wine-exposed fetuses weighed significantly less than beer-exposed fetuses (p < 0.01).

Discussion

These data corroborate previous studies demonstrating prenatal growth retardation resulting from in utero alcohol exposure [1] and extend these observations by showing that congeners present in beer and wine potentiate this ethanol-related growth retardation. Other studies from this laboratory [2, 3], however, indicate that this potentiation may no longer be present at birth or postnatally. In these latter studies, offspring exposed to beverage alcohol or ethanol weighed significantly less at birth than control offspring, but wine- and beer-exposed offspring did not weigh significantly less than ethanol-exposed offspring [2]. With respect to postnatal growth female offspring exposed to wine or ethanol still continued to weigh less than control offspring at 20 weeks of age.

Although wineexposed females weighed considerably less than ethanol-treated offspring at this age, this difference was not statistically significant. Beer- and whiskey-exposed female offspring ceased to weigh less than controls by 8 weeks of age [3].

Summary

Pregnant rats were intubated with the equivalent of 3 g/kg of ethanol twice daily in the form of beer, wine, whiskey, or ethanol. Control animals received vehicle. All animals were pair-fed to ethanol-treated dams. Dams were sacrificed on gestation day 20. Fetuses exposed to alcohol weighed less than control fetuses. Beer- and wine-exposed fetuses weighed less than ethanol-exposed fetuses.

References

1   Abel, E.L.: The fetal alcohol syndrome: behavioral teratology. Psychol. Bull. *87:* 29–50 (1980).
2   Abel, E.L.: Absence of effects of congeners on rat fetal development. Pharmacol. Biochem. Behav. (in press).
3   Abel, E.L.: Prenatal exposure to beer, wine, whiskey, and ethanol: effects on postnatal growth and food and water consumption (in submission).
4   Greizerstein, H.B.; Smith, C.M.: Development and loss of tolerance to ethanol in goldfish. J. Pharmac. exp. Ther. *187:* 391–399 (1973).
5   Greizerstein, H.B.: Congeners in alcoholic beverages (in submission).
6   Winer, B.J.: Statistical principles in experimental design; 2nd ed., pp. 464–466 (McGraw Hill, New York 1971).

E.L. Abel, PhD, Research Institute on Alcoholism, 1021 Main Street, Buffalo, NY 14203 (USA)

Fetal Malnutrition: A Possible Cause of the Fetal Alcohol Syndrome[1]

Grace W.-J. Lin[2]

Center of Alcohol Studies, Rutgers University, New Brunswick, N.J., USA

Introduction

Pregnant women who drink alcohol heavily have a high risk of giving birth to children with growth retardation, congenital malformation and mental deficiency, i.e. the Fetal alcohol syndrome [8, 10]. Animal experiments also indicate that alcohol is a teratogen [3, 9, 12]. However, the mechanism by which ethanol interferes with fetal development is still unclear.

To develop normally the fetus requires nutrients derived from the maternal circulation. Among the key nutrients for embryogenesis and fetal maturation, folates and amino acids are probably the most demanding ones. Folates are essential cofactors in synthesis of purine and pyrimidine, components of DNA and RNA, and protein synthesis requires the availability of amino acids. Since ethanol interferes with the absorption and metabolism of folates and amino acids [2, 4, 6], and since deficiency in these nutrients produces fetal malformation and impairs fetal growth [13, 15] the possibility that ethanol feeding during pregnancy might induce fetal folate and/or amino acid deficiency was studied in the rats.

---

[1] This work is supported in part by a Biomedical Research Support Grant and a grant from the Charles and Johanna Busch Memorial Fund of Rutgers University.

[2] The author is grateful to Dr. *D. Lester* for support and advice during the course of this work, and to Ms. *E. Haugen* for technical assistance.

Materials and Methods

The experiments were carried out in groups of weight-matched Sprague-Dawley female rats (Charles River, 200–250 g). On gestation day 6, one group of rats was given an ethanol-containing liquid diet ad lib while the other group of rats, serving as controls, was pair-fed with the liquid diet containing sucrose substituted isocalorically for ethanol. The feeding period was either 14 days (experiment 1; gestation day 6 through 20) or 15 days (experiment 2; gestation day 6 through 21). The liquid diets were prepared from commercial liquid food, Nutrament (vanilla imitation flavor, Mead Johnson) supplemented with casein (enzymatic hydrolysate, Sigma) and choline chloride plus ethanol or sucrose. The composition of the ethanol-containing diet, expressed in kcal/100 ml, was ethanol, 30%; protein, 20%; fat, 14.8%; and carbohydrate, 35,2%. The ethanol concentration of this diet was 4.3% (w/v). The liquid diet provided 1 kcal/ml. The composition of this diet in terms of protein, vitamins and minerals is comparable to the dietary standard recommended for pregnant rats by the Committee on Animal Nutrition of the National Research Council [11].

On gestation day 20 (experiment 1) or 21 (experiment 2), the pregnant females were food deprived for 4 h. They were then anesthetized with ether and maternal blood was drawn by cardiac puncture; fetuses and placentas were removed, blotted dry and weighed immediately. Fetuses were than decapitated and blood collected in heparinized capillary tubes; fetal brains and liver were removed and weighed. Fetal blood, brains or livers from the same litter were pooled for chemical analyses.

Tissue total folate levels, following hog kidney conjugase treatment, were determined by the microbiological method of *Bird and McGlohon* [1] using *Lactobacillus casei* (ATCC 7469) as test organism and ^{5}N-formyltetrahydrofolic acid (leucovorin; gift from Lederle Laboratories, Pearl River, N. Y.) as standard. Plasma were deproteinized by the addition of sulfosalicylic acid and the free amino acids in the protein-free filtrate were determined by an amino acid analyzer (Beckman Model 120C, two-column system).

Results were analyzed by Student's t test; differences between groups were considered statistically significant when p was less than 0.05.

Results and Discussion

Food and ethanol intakes are presented in table I. Some rats required 2–3 days of adaptation to the ethanol-containing liquid diet; thus, the average food intake of the first 5 days was slightly lower than the later periods. However, calorie intakes during the whole period of experiment were similar to or higher than the reported values for rat during gestation [11]. They averaged 271 and 253 kcal/kg body weight/day for experiment 1 and experiment 2, respectively. Overall ethanol intakes averaged 11.6 and 10.9/kg body weight/day for experiments 1 and 2, respectively.

Table I. Food and ethanol intake during gestation (mean ±SD values; number of rats in parentheses)

Gestation day	Food Intake, kcal/kg/day		Ethanol intake, g/kg/day	
	experiment 1	experiment 2	experiment 1	experiment 2
6–10	252.4 ± 31.4 (7)	220.4 ± 14.7 (6)	10.8 ± 1.4 (7)	9.5 ± 0.6 (6)
11–15	291.0 ± 22.6 (7)	271.6 ± 26.8 (6)	12.5 ± 1.0 (7)	11.6 ± 1.5 (6)
16–20	268.9 ± 24.1 (7)		11.5 ± 1.0 (7)	
16–21		265.9 ± 26.0 (6)		11.4 ± 1.1 (6)

Table II. Maternal, fetal and placental weights and litter size

Item	Experiment 1		Experiment 2	
	control	EtOH	control	EtOH
Number of dams	7	7	6	6
Maternal weight, g				
Initial	222.0 ± 10.6[1]	225.3 ± 18.5	237.8 ± 9.8	237.8 ± 17.3
Final	332.7 ± 21.7	345.4 ± 37.4	355.5 ± 22.9	344.7 ± 24.3
Gain	110.7 ± 15.7	120.1 ± 30.0	117.7 ± 15.5	106.8 ± 17.2
Litter size	10.7 ± 3.1	12.3 ± 1.3	12.7 ± 2.3	12.2 ± 1.3
Fetal weight, g	2.25 ± 0.19	2.11 ± 0.16	3.42 ± 0.29	3.09 ± 0.12[2]
Placental weight, g	0.43 ± 0.05	0.45 ± 0.07	0.47 ± 0.04	0.48 ± 0.02

[1] Mean ± SD of 7 dams or 7 litters (experiment 1), or of 6 dams or 6 litters (experiment 2).
[2] Significantly different from control ($p = 0.027$).

Table III. Brain and liver total folate contents of 20-day-old rat fetuses (experiment 1): mean ± SE, number of litters shown in parentheses

	Control, µg/g wet tissue	Ethanol, µg/g wet tissue	p value
Brain	0.87 ± 0.06 (6)	1.01 ± 0.10 (6)	0.400
Liver	8.17 ± 0.66 (7)	8.85 ± 0.41 (7)	0.397

Maternal, fetal and placental weights and litter size are presented in table II. No differences between treated and control groups were found in maternal weight gain, litter size and placental weight in either experiment. However, in both experiments, the average weight of fetus from the ethanol-treated dams was smaller than those from the controls, and the difference was statistically significant in 21-day-old fetus (3.09 vs. 3.42 g; p = 0.027) when fetal growth was at maximum.

In table III total folate contents of brain and liver of 20-day-old fetuses are presented. No statistically significant difference was found between the ethanol-treated and the control group in brain or liver; the combined means were 0.99 and 8.52 µg/g wet tissue, respectively.

Horne et al. [7] reported that ethanol stimulates the accumulation of ^{5}N-methyltetrahydrofolate in isolated rat liver cells. *Finkelstein* et al. [5] found that administration of alcohol to rats significantly decreased the activity of ^{5}N-methyltetrahydrofolate-homocysteine methyltransferase in liver. In mammals, this enzyme is essential for the generation of tetrahydrofolate from ^{5}N-methyltetrahydrofolate, an inactive storage form of folate [14]. Thus, exposure to ethanol could cause the depletion of other folate coenzymes essential for DNA and RNA syntheses. In the present study only the total folates were determined. The distribution of folate coenzymes may differ between these two groups of rats even though total folate contents are similar. Only by determining the folate coenzyme pattern can the folate status of the fetus be assessed.

Concentrations of fetal plasma free amino acids from experiment 1 are presented in table IV. The majority of the plasma-free amino acids from ethanol-treated fetus tended to be lower in comparison to those of control, but not significantly except for histidine. Histidine, an essential amino acid, was significantly lower by 41% in the ethanol-treated group as compared to the control (48.6 vs. 82.1 nmol/ml plasma; p = 0.002). Among other essential amino acids phenylalanine seems to show some degree of ethanol effect with a 24% decrease in the ethanol group. However, due to large variation within a group, the difference between the groups was found to be not significant.

Since the concentration of plasma-free histidine was greatly influenced by ethanol-feeding, and since ethanol is known to interfere with the active transport of amino acids [2], the lower histidine level found in the fetal plasma of the ethanol-fed group could be the result of poor maternal absorption or impaired placental transfer or both. To clarify this point, the free histidine concentration was determined in both maternal and fetal plasma. An amino acid analyzed was employed; other basic amino acids, arginine and lysine, were also determined.

As shown in table V, maternal plasma histidine levels did not differ between the ethanol-fed and the control group, but again fetal plasma histidine from ethanol-fed dams was greatly reduced (51%; 51.6 vs. 108.4

Table IV. Concentrations of free amino acids in fetal plasma (experiment 2): mean ± SE, number of litters are shown in parentheses

Amino acid	Control, nmol/ml plasma	Ethanol, nmol/ml plasma	p value
Alanine	1,460.3 ± 187.2 (5)	1,199.3 ± 126.6 (5)	0.370
Arginine	194.6 ± 9.9 (7)	189.6 ± 12.2 (7)	0.674
Aspartic acid	255.8 ± 48.0 (5)	176.3 ± 26.0 (5)	0.300
Glutamic acid	766.6 ± 47.3 (5)	615.2 ± 69.5 (5)	0.066
Glycine	1,000.6 ± 204.7 (5)	503.3 ± 107.8 (5)	0.128
Histidine	82.1 ± 4.4 (7)	48.6 ± 4.8 (7)	0.002
Isoleucine	158.5 ± 4.4 (5)	178.3 ± 11.4 (5)	0.277
Leucine	357.3 ± 3.8 (5)	386.5 ± 23.4 (5)	0.308
Lysine	1,043.6 ± 61.8 (7)	1,020.2 ± 42.9 (7)	0.770
Methionine	126.0 ± 8.8 (5)	109.2 ± 9.0 (5)	0.178
Phenylalanine	244.0 ± 17.1 (5)	186.3 ± 13.5 (5)	0.095
Proline	467.4 ± 53.5 (5)	424.9 ± 39.1 (5)	0.550
Serine	754.8 ± 58.2 (5)	928.5 ± 93.9 (5)	0.313
Threonine	651.9 ± 64.6 (5)	567.0 ± 45.3 (5)	0.331
Tyrosine	191.9 ± 12.6 (5)	164.6 ± 11.4 (5)	0.264
Valine	440.2 ± 10.6 (5)	468.6 ± 32.9 (5)	0.534

Table V. Free arginine, histidine and lysine levels in fetal and maternal plasma (experiment 2)

Amino acid		Fetal plasma (F), nmol/ml	Maternal plasma (M), nmol/ml	F:M ratio
Arginine	control	163.2 ± 13.6	77.4 ± 8.8	2.24 ± 0.30
	EtOH	182.2 ± 9.7	114.2 ± 19.7	1.89 ± 0.37
		NS	NS	NS
Histidine	control	105.4 ± 11.3	36.7 ± 2.1	2.96 ± 0.47
	EtOH	51.6 ± 7.0	46.7 ± 6.3	1.31 ± 0.32
		p = 0.002	NS	p = 0.015
Lysine	control	1,465.6 ± 107.7	712.0 ± 66.4	2.12 ± 0.18
	EtOH	1,220.1 ± 51.0	890.3 ± 161.1	1.61 ± 0.29
		NS	NS	NS

Results presented as means ± SE of 6 rats.
NS = No significant differences between EtOH and control values.

nmol/ml plasma; p = 0.002). No other differences in amino acid levels in these two groups were apparent in either maternal or fetal plasma. The ratio of fetal plasma amino acid to maternal plasma amino acid (F/M ratio) showed no difference between the ethanol-treated and the control groups in either arginine or lysine, but in histidine the F/M ratio was significantly lower (56%) in the ethanol-fed group as compared to the control (1.31 vs. 2.96; p = 0.015). These results suggest that the placenta is the site of ethanol action, causing fetal malnutrition in histidine.

Since maternal proteins are not transferred to the fetus in quantities of any nutritional significance, the fetus must synthesize its own proteins. The rate of protein synthesis is dependent on the concentrations of essential amino acids. Thus, deficiency in any essential amino acid in the fetal circulation would cause impairment in fetal protein synthesis and possibly produces the fetal alcohol syndrome.

Summary

The effects of ethanol ingestion during pregnancy on total folate levels in fetal tissues and on the concentrations of free amino acids in fetal and maternal plasma were examined in the rat. No differences were observed between the ehanol-fed and the control groups in total folates in fetal brain and liver. However, the concentration of fetal plasma histidine was reduced by 50% as a result of maternal ethanol consumption; the maternal plasma histidine level was not affected. It is suggested that fetal malnutrition in an essential amino acid, histidine, could impair fetal protein synthesis producing the fetal alcohol syndrome.

References

1 Bird, O.D., and McGlohon, V.M.: Differential assays of folic acid in animal tissues; in Kavanagh Analyt. microbiol., vol. 2, pp. 409–437 (Academic Press, New York 1972).
2 Chang, T.; Glazko, A.J.: Effect of ethanol on intestinal amino acid transport; in Rothschild, Oratz, Schreiber, Alcohol and abnormal protein biosynthesis: biochemical and clinical, pp. 95–110 (Pergamon Press, New York 1975).
3 Chernoff, G.F.: The fetal alcohol syndrome. An animal model. Teratology 15: 223–230 (1977).
4 Eichner, E.R.; Hillman, R.S.: Effect of alcohol on serum folate level. J. clin. Invest. 52: 584–591 (1973).
5 Finkelstein, J.D.; Cello, J.P.; Kyle, W.E.: Ethanol-induced changes in methionine metabolism in rat liver. Biochem. biophys. Res. Commun. 61: 525–531 (1974).

6   Halsted, C.H.; Griggs, R.C.; Harris, J.W.: The effect of alcoholism on the absorption of folic acid (H-PGA) evaluated by plasma levels and urine excretion. J. Lab. clin. Med. *69:* 116–131 (1967).
7   Horne, D.W.; Briggs, W.T.; Wagner, C.: Ethanol stimulates 5-methyltetrahydrofolate accumulation in isolated liver cells. Biochem. Pharmac. *27:* 2069–2074 (1978).
8   Jones, K.L.; Smith, D.W.: Recognition of the fetal alcohol syndrome in early infancy. Lancet *ii:* 999–1001 (1973).
9   Kronick, J.B.: Teratogenic effects of ethyl alcohol administered to pregnant mice. Am. J. Obstet. Gynec. *124:* 676–680 (1976).
10 Lemoine, P.; Haronsseau, H.; Borteyru, J.-P.; Menuet, J.-C.: Les enfants de parents alcooliques: anomalies observées. A propos de 127 cas. Ouest méd. *25:* 477–482 (1968).
11 National Research Council (NRC), Committee on Animal Nutrition: Nutritional requirements of laboratory animals, pp. 56–93 (National Academy of Sciences Publication, Washington 1972).
12 Randall, C.L.; Taylor, J.; Walker, D.W.: Ethanol-induced malformations in mice. Alcoholism *1:* 219–224 (1977).
13 Roe, D.A.: Fetal malnutrition, abnormal development, and growth retardation; in Drug-induced nutritional deficiencies, pp. 187–201 (AVI, Westport 1978).
14 Stokstad, E.L.R.; Koch, J.: Folic acid metabolism. Physiol. Rev. *47:* 83–116 (1967).
15 Tuchmann-Duplessis, H.: Nutrition of the embryo; in Drug effects on the fetus, pp. 20–37 (ADIS Press, Sydney 1975).

Dr. Grace W.-J. Lin, Center of Alcohol Studies, Rutgers University, New Brunswick, NJ 08903 (USA)

Fetal Alcohol Syndrome

Surendra K. Varma, Bharat B. Sharma

Department of Pediatrics, Texas Tech University Health Sciences Center School of Medicine, Lubbock, Tex., USA

Introduction

In 1968, *Lemoine* et al. [17] studying 127 children of 69 chronically alcoholic mothers, found an increased incidence of 'odd facies, growth retardation, other malformations, and psychomotor disturbances'. *Jones* et al. [11] and *Jones and Smith* [12] in 1973, described 11 unrelated children born to alcoholic women of three different ethnic groups who demonstrated similar patterns of retarded growth and development and craniofacial, limb and cardiovascular defects. This constellation of anomalies was termed the fetal alcohol syndrome.

Historical Review

The clinicians and researchers of the 18th and 19th century found the harmful effects of maternal drinking to be a topic of major concern. In 1621 *Burton* [4], in the 'Anatomy of Melancholy' cited *Aristotle's* observation that 'foolish and drunken and harebrained women most often bring forth children like unto themselves, morose and lanquid'. A committee of the Middlesex Sessions reported in 1736, 'unhappy mothers habituate themselves... children are born weak and sickly, and often look shrivel'd and old, as though they had numbered many years' [3].

With the foundation of the American Temperance Society in 1826, physicians as well as moralists threw themselves into the debate over the relative virtues of temperance (moderate drinking) and total abstinence. In 1837, *Ryan* [27] stated: 'These liquors injure the pregnant woman and expose her to danger during parturition, and to fever or inflammation afterwards, while they arrest the growth and destroy the health of the infant'. *Morel* [24] in 1857, published an elaborate theory of degeneracy which stated that parental drunkenness produced depravity, alcoholic excess and degradation in the first generation of offspring, with progressively more severe symptoms in their children, until the fourth generation developed sterility, heralding the extinction of the line.

In 1899, *Sullivan* [33] investigated female alcoholics at the Liverpool prison. He found higher rates of stillbirth and mortality under the age of 2 in the alcoholics' children, with convulsions being the most common mode of death. He correlated infant death and birth defects with conception during intoxication.

In 1910, *Elderton and Pearson* [8] concluded that 'no marked relation has been found between the intelligence, physique or disease of the offspring and parental alcoholism in any of the categories investigated'. *Ballantyne* [2] divided pregnancy into the germinal, embryonic and fetal stages, explaining that during the embryonic stage, which comprised most of the first trimester, alcohol produced structural anomalies, since organs were being formed, whereas in the fetal stage, the second and third trimester, disease or abortion would be expected as the result of heavy drinking. *Butler* [5] studied 1,438 defective delinquents, whose intelligence quotients averaged 50–70, admitted by a state home from 1931 to 1941. He found that 21% of the males and 23% of the females had a family history of alcoholism.

In the early part of the 20th century, the animal experiments which were conducted remained as inconclusive as the earlier works. *Stockard* [29], who by 1923 had been breeding this line of alcoholized guinea pigs for 13 years, continued to report reduced production of viable young and increased infant mortality in offspring of treated animals. *Mirone* [23] found in 1952 that alcohol decreased fertility of male mice and successful gestations of females. In 1962, *Schaefer* [28] reported a case of a Yukon Indian born with an apparent alcohol withdrawal syndrome. In 1970, *Ulleland* et al. [34] associated maternal alcoholism with 41% of a group of children born underweight for gestational age.

Animal Studies

In the past few years, a number of animal studies have been carried out on the effects of alcohol on the fetus. To date, there are animal models of fetal alcohol syndrome in chickens [30], mice [15], rats [22], guinea pigs [30] and zebra fish [16]. It is now clear that even in the presence of adequate nutrition during pregnancy, maternal ingestion of alcohol in experimental animals can produce growth deficiency, malformation, and increased stillbirths in the offspring [31].

Pilstrom [25] and *Wallgren and Barry* [36] reviewed the literature on fetal mortality and prenatal growth deficiency after the administration of alcohol to pregnant rats. In those investigations the fetal mortality was higher and the birth weight lower in the offspring of the alcohol-treated rats than in controls.

In experiments with pregnant sheep, the placental transport was studied and revealed a rapid placental diffusion of alcohol and a highly significant correlation between maternal and fetal blood ethanol concentrations [20].

Chesler et al. [6] investigated the relative effect of toxic doses of alcohol administered intraperitoneally in a 20% solution to rat fetuses, newborns and adult rats. Compared with adults, newborn rats showed greater resistance to the lethal effect of alcohol. LD_{50} was about 8 g/kg/24 h after the injection in newborn rats and 5 g/kg in adult rats. The corresponding values in the rat fetuses were much the same as in the adults.

The effects of ethanol on fetal cerebral function and metabolism have been studied in pregnant sheep [21]. The fetal cerebral uptake of oxygen was unaffected. The blood flow was significantly increased. The EEG showed a decrease in amplitude and slowing of the dominant rhythm as the blood ethanol concentration increased and occasionally became isoelectric, a change associated with a severe fetal acidosis during the post-infusion period.

In a study in rats, the female offspring of the mother given ethanol were more emotionally reactive than were the female offspring of control dams, when investigated by an open-field technique [1].

Rawat [26] found that 2 weeks of ethanol consumption by pregnant rats resulted in an increase in the fetal cerebral contents of gamma-aminobutyric acid and glutamate, and the cerebral activities of GABA-transferase and glutamate decarboxylase were significantly decreased in the fetal brains from ethanol-fed rats as compared with those in the controls. Ethanol feeding to the pregnant rats did not result in a change in the steady state concentration of 5-hydroxytryptamine in the fetal brain. Increased levels of noradrenaline were observed in both the fetal and neonatal brains from ethanol-fed groups as compared with the noradrenaline content in the corresponding controls.

Studies in Humans

Fuchs et al. [9] showed that ethanol administered to pregnant women may prevent threatened premature labor. If this treatment is unsuccessful the blood of the infant has been found to contain alcohol at birth. The blood alcohol level at birth has been reported to range from 0.4 to 1.83 g/l. No significant CNS depression was observed or any noteworthy changes in alertness, motor activity, circulation or acid-base status. *Wagner* et al. [35], who infused alcohol directly into infants for experimental purposes, reported that 5 of 6 infants fell asleep 1 h after the infusion which they ascribed to a mild sedative effect of alcohol.

Lopez and Montoya [19] observed vacuolization of cells in early erythropoiesis and myelopoiesis in 2 premature infants whose mother had received alcohol ad modum. The changes disappeared in one month. This condition is also observed in adults after heavy drinking but it usually disappears within 2 weeks.

Jones et al. [13] studied the outcome in offspring of chronic alcoholic mothers. They found 23 children by the mother's history of alcoholism and examined 13 of them aged 7 years or more. 6 children could not be traced and 4 children had died in the neonatal period, a 17% neonatal mortality rate. Each of the 13 was matched with 2 controls. They again found over-representation of the structural anomalies.

Streissguth [32] reported on the mental development of 12 children born to alcoholic mothers, relating the severity of their mental deficiency to the severity of the physical characteristics of the fetal alcohol syndrome. Their intellectual deficit seemed to be the result of brain damage occurring in utero.

Little [18] surveyed 811 middle-class pregnant women and found that even moderate alcohol use during pregnancy was associated with decreased birth weight.

Autopsy data is available on 1 case [13]. The patient was a preterm infant born at 32 weeks gestation, who died following a series of apneic spells. Multiple heteropias, fusion of the anterior superior gyri, lissencephaly and agenesis of the corpus callosum were found in the brain, which weighed only 140 g.

Clarren et al [7] have reported neuropathological observations on brains of infants born to women who drank heavily during pregnancy. These authors evaluated 14 brains of human infants exposed in utero to variable quantities of alcohol. Brain malformations were found in 5 of 6 infants whose mothers at least occasionally consumed five or more drinks at a time during pregnancy. The brain malformations observed consisted of severe disturbances of neuronal migration.

Social Drinking

Kaminsky et al [14] in 1976 published a study on 9,000 pregnant women in France and showed significantly decreased birth weight in offsprings of mothers ingesting over 1.6 ounces of absolute alcohol daily even when other risk factors associated with higher alcohol use were controlled for. *Little* [18] reported a similar finding in a study done in Seattle on several hundred middleclass pregnant women in a prepaid medical program. Maternal alcohol use during the 6 months prior to pregnancy and during the eighth month of pregnancy were significantly related to birthweight of offspring. The critical level of alcohol was 1 ounce of absolute alcohol consumed on an average daily basis.

The effects of moderate levels of alcohol consumption are not definitive. Consumption levels commonly called social drinking are significantly related to decreased birthweight in the offspring, as well as to a variety of behavioral deficits of unknown predictability.

Etiology

The etiological mechanism is unclear. If, as seems likely, the cause is direct toxicity, then ethanol or its metabolite, e.g., acetaldehyde are likely candidates. There is good evidence in humans and other animals that ethanol freely crosses the placental barrier [20].

Clinical Features

Children with fetal alcohol syndrome have a characteristic facies with short palpebral fissures as the most differentiating feature, and often have a flattened nasal bridge and epicanthic folds. Midfacial hypoplasia, microphthalmos, intraocular defects, strabismus, ptosis of eyelids and cleft palate are other less frequent anomalies [10].

Elsewhere in the body, skeletal anomalies (clinodactyly, camptodactyly and mild joint restrictions), cardiac malformations (primarily atrial septal defect, followed by ventricular septal defect), abnormal palmar creases and, in girls, hypoplasia of the labia majora are found.

Mental deficiency, ranging from borderline to severe, has been found in most of such children.

There is severe intrauterine growth retardation. The infants are both short and light for dates, and the growth potential in childhood is poor. Almost all children have microcephaly.

The affected children often fail to thrive in terms of survival, neonatal adaptation, brain function, and growth. There is an increased perinatal mortality. Tremulousness, hyperactivity, and irritability are variable features during the neonatal period. Tremulousness frequently persists for months and even years, while the fine motor dysfunction and development delay may be permanent. The growth rate is irreversibly reduced, since the infants placed in foster homes have remained small and mentally defective. In the patients who could be followed after 1 year of age, linear growth rate was 65% of normal, while the average rate of weight gain was only 38% of normal.

Conclusion

Current data clearly point to an association between chronic maternal alcoholism and serious morphological and developmental abnormalities in the fetus. Women who are alcoholics should know the risk of the alcoholism giving rise to a serious problem in the developing fetus. This risk is estimated to be between 30 and 50%. Ideally, such women should be encouraged and assisted in exercising effective birth control until such time as they can discontinue the alcohol intake.

Summary

A historical review of alcohol drinking and its effect on fetus, and animal and human studies on the fetal alcohol syndrome are described. The clinical features of this syndrome include characteristic facies with short palpebral fissures, skeletal anomalies, cardiac malformations, mental deficiency and growth retardation. The review presented concludes that there is clear association between chronic maternal alcoholism and serious morphological and developmental abnormalities in the fetus.

References

1 Abel, E.L.: Emotionality in offspring of rats fed alcohol while nursing J. Stud. Alcohol *36:* 654–658 (1975).
2 Ballantyne, J.W.: The pathology of antenatal life. Glasgow med. J. *49:* 241–258 (1898).
3 Beecher, L.: Six sermons on the nature, occasions, signs, evils, and remedy of intemperance (Crocker & Brewter, Boston 1927).
4 Burton, R.: The anatomy of melancholy, vol. I, part I, section 2. Causes of melancholy (William Tegg, London 1906).
5 Butler, F.O.: The defective delinquent. Am. J. ment. Defic. *47:* 7–13 (1942).
6 Chesler, A.; La Belle, G.C.; Himwich, H.E.: The relative effects of toxic doses of alcohol on fetal, newborn and adult rats. Q. Jl Stud. Alcohol *3:* 1–4 (1942).
7 Clarren, S.K.; Alvord, E.C.; Sumi, S.M.; Steissguth, A.P.: Brain malformations related to prenatal exposure to ethanol. J. Pediat. *92:* 64–67 (1978).
8 Elderton, E.; Pearson, K.: A first study of the influence of parental alcoholism on the physique and ability of the offspring. Eugenics laboratory memoir X (Cambridge University Press, London 1910).
9 Fuchs, F.; Fuchs, A.R.; Poblete, V.R.; Riik, A.: Effect of alcohol on threatened premature labor. Am J. Obstet. Gynec. *99:* 627–637 (1967).
10 Hanson, J.W.; Jones, K.L.; Smith, D.W.: Fetal alcohol syndrome. Experience with 41 patients. J. Am. med. Ass. *235:* 1458–1460 (1976).
11 Jones, K.L.; Smith, D.W.; Ulleland, C.N.; Streissguth, A.P.: Pattern of malformation in offspring of chronic alcoholic mothers. Lancet *i:* 1267–1271 (1973).
12 Jones, K.L.; Smith, D.W.: Recognition of fetal alcohol syndrome in early infancy. Lancet *ii:* 999 (1973).
13 Jones, K.L.; Smith, D.W.; Streissguth, A.P.; Myrianthyopoulos, N.C.: Outcome in offspring of chronic alcoholic women. Lancet *i:* 1076–1078 (1974).
14 Kaminsky, M.; Rumeau-Rouquette, C.; Schwartz, D.: Consommation d'alcool chez les femmes enceintes et issue de la grossesse. Revue Epidémiol. Méd. soc. Santé publ. *24:* 27–40 (1976).
15 Kronick, J.: Teratogenic effects of ethyl alcohol administered to pregnant mice. Am. J. Obstet. Gynec. *124:* 676–680 (1976).
16 Laale, H.: Ethanol induced notochord and spinal cord duplications in the embryo of the zebra fish, *Brachydanio rerio.* Expl Zool. *177:* 51–64 (1971).
17 Lemoine, P.; Harousseau, H.; Borteyru, J.P.; Menuet, J.C.: Les enfants des parents alcooliques: anomalies observées. A propos de 127 cas. Ouest méd. *21:* 476–482 (1968).
18 Little, R.E.: Moderate alcohol use during pregnancy and decreased infant birth weight. Am. J. Publ. Hlth *67:* 1154–1156 (1977).
19 Lopez, R.; Montoya, M.R.: Abnormal bone marrow morphology in the premature infant associated with maternal alcohol infusion. J. Pediat. *79:* 1008–1010 (1971).
20 Mann, L.I.; Bhakthavathsalan, A.; Liu, M.; Makowski, P: Placental transport of alcohol and its effect on maternal and fetal acid-base balance. Am. J. Obstet. Gynec. *122:* 837–844 (1975).
21 Mann, L.I.; Bhakthavathsalan, A.; Liu, M.; Makowski, P.: Effect of alcohol on fetal cerebral function and metabolism. Am. J. Obstet. Gynec. *122:* 845–851 (1975).

22 Martin, J.: Offspring survival and development following maternal ethanol administration. Devl. Psychobiol. *10:* 5 (1977).
23 Mirone, L.: The effect of ethyl alcohol on growth, fecundity and voluntary comsumption of alcohol by mice. Q. Jl. Stud. Alcohol *13:* 365–369 (1952).
24 Morel, B. A.: Traité des dégénérescences physiques, intellectuelles, et morales de l'espèce humaine et des causes qui produisent ces variétés maladives (Baillière, Paris 1857).
25 Pilstrom, L.: Some effects of ethanol intake on cellular metabolism and subcellular structure. Acta univ. uppsaliensis, Vol. 170 (1970).
26 Rawat, A.: Effects of maternal ethanol consumption on the fetal and neonatal cerebral neurotransmitters; in Lindros, Eriksson, the role of acetaldehyde in the actions of ethanol. Satellite Symp. Sixth Int. Congr. Pharmacology, Helsinki 1975, vol. 23, pp. 159–176 (Finnish Foundation for Alcohol Studies, Helsinki 1975).
27 Ryan, M.: The philosophy of marriage in its social, moral and physical relations (Churchill, London 1837).
28 Schaefer, O.: Alcohol withdrawl syndrome in a newborn infant of a Yukon Indian mother. Can. med. Ass. J. *87:* 1333–1334 (1962).
29 Stockard, C. F.: Experimental modifications of the germ-plasm and its bearing on the inheritance of acquired characters. Proc. Am. phil. Soc. *62:* 311–325 (1923).
30 Streissguth, A.: Maternal alcoholism and the outcome of pregnancy. A review of the fetal alcohol syndrome; in Alcohol problems in women and children (Grune & Stratton, New York 1976).
31 Streissguth, A. P.: Maternal drinking and the outcome of pregnancy: implications for child mental health. Am. J. Orthopsychiat. *47:* 422–431 (1977).
32 Streissguth, A. P.: Psychological handicaps in children with fetal alcohol syndrome. Ann. N. Y. Acad. Sci.
33 Sullivan, W. C.: A note on the influence of maternal inebriety on the offspring. J. ment. Sci. *45:* 489–503 (1899).
34 Ulleland, C.; Wennberg, R. P.; Igo, R. P.; Smith, N. J.: The offspring of alcoholic mothers. Pediat. Res. *4:* 474 (1970).
35 Wagner, L.; Wagner, G.; Guerrero, J.: Effect of alcohol on premature newborn infants. Am. J. Obstet. Gynec. *108:* 308–315 (1970).
36 Wallgren, H.; Barry, H.: Actions of alcohol, vol. 1, 2 (Elsevier, Amsterdam 1970).

S. K. Varma, M. D., Department of Pediatrics, Texas Tech University Health Sciences Center School of Medicine, Lubbock, TX 79430 (USA)

Effects of Sex Steroids on Ethanol Pharmacokinetics and Autonomic Reactivity

Arthur R. Zeiner, Pamela S. Kegg

University of Oklahoma Health Sciences Center, Oklahoma City, Okla., USA

Several converging lines of evidence suggest that sex steroids may be involved in modulating ad lib alcohol consumption, ethanol pharmacokinetics and acetaldehyde pharmacokinetics. Of concern and interest are both the effects of ethanol on endocrine function and its converse, the effects of endocrines on ethanol pharmacokinetics. The former question has had several competent reviews [31, 36, 47, 59, 62]. The effects of ethanol on endocrine function in males will not be further discussed in the present review other than to note that a sizeable data base exists with both animals and humans which suggests that chronic ethanol ingestion leads to hypogonadism, estrogenization and feminization of males [5–8, 10, 15, 21, 44–46, 56, 64–66]. Recent evidence suggests that ethanol may be a gonadal toxin for both sexes [20]. *Cicero* et al. [9] have shown that acetaldehyde, the first metabolite of ethanol metabolism, is an even more potent inhibitor of testicular steroidogenesis than is ethanol.

Females, on the other hand, have been investigated very little or not at all with respect to either the first or second question. Almost all of the research as well as treatment programs for alcoholism have been carried out on males with the tacit assumption that results from males would generalize directly to females. It is just now being realized by policy makers and experts that results based on males may not be directly applicable to females. The present review focuses on variation in ethanol pharmacokinetics as a function of variation of sex steroids in females.

Researchers have adopted several strategies to investigate the relationship of sex hormones to drug action. At the most basic level ovariectomized animals are used and then given various doses of supplemental hormones. Animal data are both useful and suggestive but they are not definitive in predicting reactions of human females. Of ultimate interest is what happens to humans. Although it is not possible to exert the types of controls feasible in animal experiments, approximations can be made. Thus, in human studies it is possible to select females on birth control pills (high sex steroid concentrations) and compare their ethanol pharmacokinetics to females not on birth control pills (lower sex steroid concentrations). Alternately, pregnant females can be compared with nonpregnant females. Another approach is to compare ovariectomized patients not on hormone supplements with normally cycling females. Finally, scientists can utilize one of the experiments of nature, the fact that sex steroids as well as estrogen/progesterone ratios vary in a lawful fashion over the menstrual cycle. At one time or other all of the above strategies have been utilized.

Animal Studies

Several experiments have demonstrated that rats will decrease their voluntary ethanol consumption after receiving estrogen injections [2–4]. Further, during the estrous stage at a time of elevated estrogen concentration, rats will drink less alcohol than at other times [3, 4]. *Mardones* [42] noted that diethylstilbesterol significantly reduced ethanol intake in either normal or gonadectomized males and females, whereas no such effects were demonstrable with testosterone. *Emerson* et al. [16] found that high doses of estradiol benzoate reduced ethanol consumption in deer mice whereas progesterone, even in massive doses, did not alter ethanol preference in the same species. *Ericksson* [17], tested for the effects of ovariectomy and contraceptive hormones on voluntary alcohol consumption in 3-month-old female rats. He demonstrated that before ovariectomy experimental and control animals did not differ on ad lib alcohol intake, after ovariectomy there was a small decrease in ethanol intake, and contraceptive hormones had a strong inhibiting effect on ethanol consumption of both ovariectomized and control animal.

Animal studies indicate that several steroids are competitive inhibitors of ethylmorphine, hexobarbital and zoxa-zolamine metabolism in vitro [29, 60, 61]. *Jori* et al. [28] demonstrated that the converse also holds. Sex steroids can also act as inducers to increase metabolism of *p*-nitroamisol aminopyrine and aniline in vitro. *Einarsson* et al. [14] demonstrated that estrogenic compounds given in higher than physiological doses have a suppressing effect on microsomal steroid metabolizing activities.

Human Studies

The effect of menstrual cycles on behavoir has been a fertile field for both speculation and research for some time [55]. Prior to an interest in sex steroid ethanol interactions, menstrual cycle effects have been related to mood and alertness [51], measures of arousal [35, 68], visual sensitivity and sexual arousal [13], cognitive perceptual behavior [6, 58], and autonomic changes [19, 39, 70, 71, 72, 73]. It has also been shown that in humans, female-initiated sexual activity peaks around the point of ovulation [1]. This activity is suppressed in females using oral contraceptives.

While research into mood, performance and behavior correlates of the menstrual cycle has been going on for some time, it has only recently been recognized that there may be relationships between behavior or feeling states, endocrine functions and alcohol.

It is known that several changes take place over the 28 days of the normal menstrual cycle in females who are not on birth control pills [22, 57]. Starting with day 1, the first day of menstrual flow, both estradiol and progesterone concentrations are at their lowest point and remain low over the first 4 days. Then estradiol concentrations start to increase and peak at about day 14, around the time of ovulation. Following ovulation estradiol shows a transitory decrease and then both estradiol and progesterone increase together until about days 24–26. At this time both progesterone and estradiol concentrations drop dramatically and this starts the onset of the next period. This little experiment of nature provides the basis for many of the alcohol-menstrual cycle effect interaction studies. Since both estrogen and progesterone vary over the menstrual cycle, judicious selection of times during the cycle makes it possible to manipulate hormones indirectly and observe the effects of this variation on ethanol pharmacokinetics. Thus, for example, selection of days 1–4 of the menstrual cycle for testing should yield very low concentrations of both estradiol and progesterone; selection of midcycle (day 14) should yield high estradiol and low progesterone concentrations; and selection of testing times around days 24–26 should yield high estradiol and high progesterone concentrations. Other ways of more directly manipulating these hormones are by comparison of groups on birth control pills versus groups of women with normal menstrual cycles.

Sex Steroid-Drug Interaction

The effect of drug interaction with oral contraceptive steroids has become a topic of concern and research. Of interest here is whether sex steroids might interfere with concomitantly administered drugs. *O'Malley* et al. [50] demonstrated that in human females on oral contraceptive steroids, the drug-metabolizing capacity for antipyrine was impaired when compared with controls not on oral contraceptives.

Crawford and Rudofsky [12] performed a series of illuminating experiments where they tested pregnant females, neonates, patients with hystero-oophorectomy, postmenopausal patients not on supportive hormone therapy, postmenopausal patients on stilbesterol, subjects taking oral contraceptives and male controls. The excretion of pethidine and promazine was studied. Results indicated that (1) pregnancy is associated with a decreased ability to metabolize certain drugs; (2) administration of oral contraceptives promotes a similar deficiency; (3) a newly born infant shows the deficiency of the mother; (4) stilbesterol administration leads to a diminution of metabolizing capacity; (5) progesterone administration to males leads to a diminished capacity to metabolize drugs, and (6) an effect of ovary removal on drug metabolism was not demonstrated.

Neims et al. [49] and *Kling and Christensen* [34] demonstrated that caffeine elimination is prolonged during pregnancy especially in its later stages. *Patwardhan* et al. [52] tested for caffeine elimination in males, females on oral contraceptive steroids and females not on oral contraceptive steroids. Females on oral contraceptive steroids took significantly longer to eliminate caffeine than did males or females not on oral contraceptive steroids. The latter two groups did not differ from each other. Data of several studies reviewed suggest that oral contraceptive steroids impair elimination of a variety of drugs.

Epidemological Data on Sex Steroid-Alcohol Interactions

Little et al. [40] noted that there was a sharp decrease in consumption of alcoholic beverages during pregnancy. Patients cited adverse physiological effects as the major reason for the decline in drinking during pregnancy. In a follow-up study, *Little and Streissguth* [41] demonstrated that decreased alcohol intake during pregnancy applies to alcoholic women as well. *Jones* et al. [26] also found that women taking oral contraceptives drank significantly less alcohol than did control females not on oral contraceptives. *Zeiner and Farris* [74] extended the generality of these findings by testing native American Indian females on oral contraceptive steroids and a group not on oral contraceptive steroids. Groups did not differ in age, education, socioeconomic level, height or weight. However, the group on oral contraceptive steroids demonstrated a significantly lower alcohol intake than the controls.

In summary, the studies reviewed suggest that altered hormone levels induced either by pregnancy or via oral contraceptive steroids, either take away the desire to drink alcohol or make it aversive to do so.

At another level, *Wilsnack* [69], *Jung and Russfield* [30], *Lloyd and Williams* [38], *Pincus* et al. [53] and *Ratnoff and Patek* [54] have adduced evidence for the view that the incidence of obstetrical and gynecological disorders is significantly higher among alcoholic females than among females who are not alcoholic. From their data, it is not clear whether obstetric gynecological problems are predisposing factors for alcoholism or whether alcoholism leads to obstetric and gynecological problems. The medical literature suggests that the latter may be the case.

Menstrual Cycle-Ethanol Pharmacokinetic Interactions

Jones and Jones [24, 25] tested 20 women twice during their menstrual cycle (either during menstruation, day 14, or the day preceding the first day of flow). 10 males were also tested twice at about a 2-week interval. Subjects were tested at 1:00 p.m. on a 4-hour empty stomach. All received 0.52 g/kg ethanol in a 1:4 orange drink mix which was consumed over a paced 5-min interval. They found that women reached a significantly higher peak blood alcohol concentration (BAC) than men. Highest peak BAC for the women was obtained at the premenstrual time (day before flow onset).

Jones et al. [26] demonstrated in a second experiment that females reach a higher peak BAC than do males for the same dose per unit of body weight.

Jones and Jones [25] performed another experiment to test for dose-response differences at three different doses on 2 males and females. Gender differences found at the single dose in peak BAC in their earlier experiment were maintained at different doses.

Kegg and Zeiner [32] tested for birth control pill effects on ethanol pharmacokinetics, acetaldehyde and cardiovascular measures in Caucasian females. Two groups of 7 females each either on oral contraceptive steroids or not on oral contraceptive steroids were tested twice on day 1, the first day of menstrual flow, and day 26 of the menstrual cycle. As far as possible, order of testing was counterbalanced in each group. Subjects received 0.52 g/kg 190 proof ethanol in a 1:4 orange drink after a 4-hour fast. The dose was consumed over a paced 5-min period. Subjects in the oral contraceptive group had been on pills for at least 3 months. Subjects in the control group not on oral contraceptive steroids had regular cycles of 28 ± 2 days and had not been on oral contraceptive steroids or other medications for at least 3 months. Blood alcohol and acetaldehyde concentrations were determined from breath samples via a gas chromatograph every 10 min. Cardiovascular measures of systolic, diastolic, mean arterial pressure and heart rate were recorded noninvasively before each breath sample with a Dinamap.

In contrast to findings by *Jones and Jones* [25], subjects reached a reliably higher peak BAC on day 1, at a time of low estrogen and progesterone than day 26, a time of high estrogen and progesterone. Acetaldehyde concentration at peak BAC was significantly higher on day 26 than on day 1. Acetaldehyde concentration increased reliably more from day 1 to day 26 in the group on oral contraceptives than in the control group not on oral contraceptives. The oral contraceptive groups had a reliably higher diastolic blood pressure during both baseline a well as peak BAC measures than did the control group not on oral contraceptives. Significant differences between groups were not found with systolic blood pressure, mean arterial pressure or heart rate.

Zeiner and Kegg [75] in a follow-up experiment further investigated the relationship between oral contraceptives, day of menstrual cycle and alcohol pharmacokinetics in 20 Caucasian females. Groups did not differ in age, weight, height, lean body mass, education or alcohol consumption. Subjects in the oral contraceptive group had been on birth control pills for an average of 2.6 years, controls not on oral contraceptives had regular cycles of 28±2 days. Alcohol dose, concentration, timed administration and methodology were the same as in the preceding study. Subjects were tested twice (on day 1 and day 24). Order of testing in each group was counterbalanced. They were instructed not to take any drugs including alcohol for at least 24 h before coming to the laboratory for testing and to abstain from eating or drinking for at least 4 h. A baseline reading was taken with the gas chromatograph to assure that they did in fact have zero BAC concentrations and to make sure that subjects knew the procedure. Subjects then drank their 0.52 g/kg ethanol-orange drink over a paced 5-min period. Breath samples were taken every 10 min until they returned to a BAC of 20 mg% or less.

The control group *not* on oral contraceptive steroids reached a reliably higher peak BAC than did the group on oral contraceptives. For both groups peak blood alcohol concentrations were significantly higher on day 1 when hormone levels are low than on day 24 when both estrogen and progesterone concentrations are elevated. These data replicate and extend the findings of *Kegg and Zeiner* [32].

Several good reviews [11, 48] suggest that the most likely site of action of sex steroids and/or drugs with alcohol pharmacokinetics might be with the alcohol dehydrogenase system. A less significant site, but one that could also result in reliable effects, would be via alterations in the microsomal oxidizing system. To our knowledge, definitive answers on those points do not exist and await further research.

Menstrual Cycle-Ethanol Performance Interactions

The interaction of performance measures with menstrual cycles and/or drugs such as alcohol is a controversial topic. There is some evidence that there are sensory threshold variations related to the menstrual cycle [33, 63, 67], but it is not known whether these would relate to performance differences and/or interact with alcohol.

Jones and Jones [25] did not find performance differences as a function of menstrual cycle on either immediate recall or delayed recall tasks. In a later study [23, 27], they compared menstruating and nonmenstruating alcoholic females with menstruating and nonmenstruating controls on both tasks. They concluded that cognitive deficits result from prolonged alcohol intake. A very recent experiment by *Fabian* et al.

[18] could not replicate the findings of *Jones* et al. [27]. Contrary to the findings of *Hatcher* et al. [23], there were no differences in the performances of menstruating and nonmenstruating women on any of the cognitive tests in either alcoholic or social drinker groups.

Linnoila et al. [37] studied the effects of alcohol on psychomotor performance of women and its interaction with the menstrual cycle. 10 young females not on oral contraceptives were tested with placebo and three doses of ethanol administered double-blind. Performance measures included simple reaction time to a light flash and a continuous tracking task at three levels of instability. Over eight menstrual cycles subjects were tested (once per cycle) in random order during follicular and luteal phases of the menstrual cycle. There was not a significant effect of blood alcohol concentration as a function of the menstrual cycle. A significant positive correlation was observed between blood alcohol concentration and reaction time impairment, but only during the luteal phase of the menstrual cycle. Further, these investigators found that low doses of alcohol affected performance more seriously than did higher doses.

Discussion

Several lines of evidence suggest that sex steroids exert an effect on ethanol pharmacokinetics. The evidence is most consistent and strongest in suggesting that female sex steroids, particularly estrogen, inhibit ethanol ingestion as well as perhaps pharmacokinetics. The evidence is strongest with estrogen, weaker with progesterone and contradictory with testosterone.

Thus, it has been shown in both animal models and humans that manipulation of estrogen concentrations either via contraceptive steroids, pregnancy or timed administration at certain points in the menstrual cycle, leads to decreased ethanol intake.

The proposed mechanism of action for these effects is at present not clear. It has been proposed [43] that the sex steroids may interfere with acetaldehyde metabolism and that this may be aversive in similar fashion to the alcohol-disulfiram reaction where acetaldehyde is also built up and leads to aversive consequences. *Kegg and Zeiner* [32] have evidence which is consonant with such a hypothesis. They found higher acetaldehyde concentrations after an acute dose of ethanol around day 26 of the menstrual cycle when both estrogen and progesterone concentrations are high, than on day 1 when both estrogen and progesterone concentrations are low.

Summary

Several lines of evidence indicate that female sex steroids, especially estrogen, interact with ethanol pharmacokinetics. Both animal and human studies are consistent in indicating that voluntary intake of alcohol decreases with increasing estrogen concentrations. Such effects are equivocal with progesterone and nonexistent with testosterone. There is some suggestion that estrogen modulates monoamine oxidase activity and via that mechanism alters sympathetic nervous system tone. Alterations in sympathetic nervous system tone, in turn, can alter gastric motility and absorption of alcohol from stomach and intestines. Several types of studies, both animal and human, indicate that high estrogen concentrations also impair ethanol metabolism. At present the exact mechanism of action for this effect is not known. Future studies investigating this aspect are indicated.

At present the effects of sex steroids on performance and its interaction with alcohol do not show a clear relationship. There are contrary pieces of information both for and against the view that menstrual cycle, performance and alcohol interact.

References

1 Adams, D.B.; Gold, A.R.; Burt, A.D.: Rise in female-initiated sexual activity at ovulation and its suppression by oral contraceptives. New Engl. J. Med. *299:* 1145 to 1150 (1978).
2 Aschkenasy-Lelu, P.: Action d'un œstrogène sur la consommation spontanée d'une boisson alcoolisée chez le rat. C.r. hebd. Séanc. Acad. Sci. Paris *247:* 1044–1047 (1958).
3 Aschkenasy-Lelu, P.: Relation entre l'effet inhibiteur des œstrogènes sur la consommation d'alcool du rat et leur action génitale. Archs Sci. physiol. *14:* 165–175 (1960).
4 Aschkenasy-Lelu, P.: Action inhibitrice d'une hormone œstrogène sur la consommation spontanée d'alcool par le rat. Revue fr. Etud. clin. Biol. *5:* 132–138 (1960).
5 Badr, F.; Bartke, A.: Effect of ethyl alcohol on plasma testosterone level in mice. Steroids *23:* 921–928 (1974).
6 Cicero, T.J.: Common mechanisms underlying the effects of ethanol and the narcotics on neuroendocrine function; in Mello, Advances in substance abuse. Behavioral and biological research (JAI Press, Greenwich, in press).
7 Cicero, T.J.; Badger, T.M.: A comparative analysis of the effects of narcotics, alcohol and the barbiturates on the hypothalamic-pituitary-gonadal axis. Adv. exp. Med. Biol. *85-B:* 95–115 (1977).
8 Cicero, T.J.; Badger, T.M.: Effects of alcohol on the hypothalamic-pituitary-gonadal axis in the male rat. J. Pharmac. exp. Ther. *201:* 427–433 (1977).

9 Cicero, T.J.; Bell, R.D.; Meyer, E.R.; Badger, T.M.: Ethanol and acetaldehyde directly inhibit testicular steroidogenesis. J. Pharmac. exp. Ther. *213:* 228–233 (1980).
10 Cicero, T.J.; Meyer, E.R.; Bell, R.D.: Effects of ethanol on the hypothalamic-pituitary-luteinizing hormone axis and testicular steroidogenesis. J. Pharmac. exp. Ther. *208:* 210–215 (1979).
11 Cohen, S.: The pharmacology of alcohol. Postgrad. Med. *64:* 97–102 (1978).
12 Crawford, J.S.; Rudofsky, S.: Some alterations in the pattern of drug metabolism associated with pregnancy oral contraceptives, and the newly born. Br. J. Anaesth. *38:* 445–454 (1966).
13 Diamond, M.; Diamond, L.; Mast, M.: Visual sensitivity and sexual arousal levels during the menstrual cycle. J. nerv. ment. Dis. *155:* 179–186 (1972).
14 Einarsson, K.; Gustafsson, J.A.; Sjovall, J.; Zietz, E.: Dose-dependent effects of ethinyloestradiol, diethylstilbesterol and oestradiol on the metabolism of 4-androstene-3,17-dione in rat liver microsomes. Acta endocr. Copenh. *78:* 54–64 (1975).
15 Ellingboe, J.; Varanelli, C.C.: Ethanol inhibits testosterone biosynthesis by direct action on Leydig cells. Res. Commun. chem. Pathol. Pharmacol. *24:* 87–102 (1979).
16 Emerson, G.A.; Brown, R.C.; Nash, J.B.; Moore, W.T.: Species variation in preference for alcohol and in effects of diet or drugs on this preference, J. Pharmac. exp. Ther. *106:* 384 (1952).
17 Ericksson, K.: Effects of ovariectomy and contraceptive hormones on voluntary alcohol consumption in the albino rat. Jap. J. Alcohol Stud. *4:* 1–5 (1969).
18 Fabian, M.; Hochla, N.; Silberstein, J.; Parsons, O.A.: Menstrual states and neuropsychological functioning. Biol. Psychol. Bull. *6:* 37–45 (1980).
19 Friedman, J.; Meares, R.A.: The menstrual cycle and habituation. Psychosom. Med. *41:* 369–381 (1979).
20 Gavaler, J.S.; Van Thiel, D.H.; Lester, R.: Ethanol. A gonadal toxin in the mature rat of both sexes. Alcohol. clin. exp. Res. *4:* 271–276 (1980).
21 Gordon, G.G.; Altman, K.; Southren, A.L.; Lieber, C.S.: Effect of alcohol (ethanol) administration on sex-hormone metabolism in normal men. New Engl. J. Med. *295:* 793–797 (1976).
22 Guyton, A.C.: Medical physiology (Saunders, Philadelphia 1976).
23 Hatcher, E.M.; Jones, B.M.; Jones, M.K.: Cognitive deficits in alcoholic women. Alcohol. clin. exp. Res. *1:* 371–377 (1977).
24 Jones, B.M.; Jones, M.K.: Alcohol effects in women during the menstrual cycle. Ann. N.Y. Acad. Sci. *273:* 567–587 (1976).
25 Jones, B.M.; Jones, M.K.: Intoxication, metabolism, and the menstrual cycle; in Greenblatt, Schuckit, Alcoholism problems in women and children (Grune & Stratton, New York 1976).
26 Jones, B.M.; Jones, M.K.; Paredes, A.: Oral contraceptives and ethanol metabolism. Revta invest. Clin., Mexico *28:* 95 (1976).
27 Jones, B.M.; Jones, M.K.; Hatcher, E.M.: Cognitive deficits in women alcoholics as a function of gynecological status. J. Stud. Alcohol *41:* 140–146 (1980).
28 Jori, A.; Bianchetti, A.; Prestini, P.E.: Effect of contraceptive agents on drug metabolism. Eur. J. Pharmacol. *7:* 196–200 (1969).
29 Juchau, M.R.; Fouts, J.R.: Effects of norethynodrel and progesterone on hepatic microsomal drug metabolizing enzyme systems. Biochem. Pharmac. *15:* 891 (1966).

30   Jung, Y.; Russfield, A.B.: Prolactin cells in the hypophysis of cirrhotic patients. Arch. Path. *94:* 265–269 (1972).
31   Kakihana, R.; Butte, J.C.: Ethanol and endocrine function; in Majchrowicz, Noble, Biochemistry and pharmacology of ethanol, vol. II, pp. 147–164 (Plenum Press, New York 1979).
32   Kegg, P.S.; Zeiner, A.R.: Birth control pill effects on ethanol pharmacokinetics acetaldehyde, and cardiovascular measures in Caucasian females. Psychophysiology *17:* 294 (1980).
33   Kenshalo, D.R.: Psychophysical studies of human temperature sensitivity; in Neff, Contributions to sensory physiology, vol. IV (Academic Press, New York 1970).
34   Kling, O.R.; Christensen, H.D.: Caffeine elimination in late pregnancy. Fed. Proc. *38:* 266 (1979).
35   Koppel, B.S.; Lunde, D.T.; Clayton, R.B.; Moos, R.H.: Variations in some measures of arousal during the menstrual cycle. J. nerv. ment. Dis. *148:* 180–187 (1969).
36   Lester, R.; Van Thiel, D.H.; Eagan, P.K.; Imhoff, A.F.; Fisher, S.E.: Hypothesis concerning the effects of dietary nonsteroidal estrogen on the feminization of male alcoholics; in Li, Schenker, Lumeng, Alcohol and nutrition, Res. Monogr., #II, (HEW-NIAAA, 1979).
37   Linnoila, M.; Erwin, C.W.; Ramm, D.; Cleveland, W.P.; Brendle, A.: Effects of alcohol on psychomotor performance of women. Interaction with menstrual cycle. Alcohol. clin. exp. Res. *4:* 302–305 (1980).
38   Lloyd, C.W.; Williams, R.H.: Endocrine changes associated with Laennec's cirrhosis of the liver. Am. J. Med. *4:* 315–330 (1948).
39   Little, B.C.; Zahn, T.P.: Changes in mood and autonomic functioning during the menstrual cycle. Psychophysiology *11:* 579–590 (1974).
40   Little, R.E.; Schultz, F.A.; Mandell, W.: Drinking during pregnancy. J. Stud. Alcohol *37:* 375–379 (1976).
41   Little, R.E.; Streissguth, A.P.: Drinking during pregnancy in alcoholic women. Alcohol. clin. exp. Res. *2:* 179–183 (1978).
42   Mardones, J.: Experimentally induced changes in the free selection of ethanol. Int. Rev. Neurobiol. *2:* 41–76 (1960).
43   Maxwell, E.S.; Topper, Y.J.: Steroid sensitive aldehyde dehydrogenase from rabbit liver. J. biol. Chem. *236:* 1032–1037 (1961).
44   Mendelson, J.H.; Mello, N.K.: Alcohol, aggression and androgens; in Frazier, Aggression. Proc. Asso. Res. Nervous and Mental Disease, pp. 25–247 (Williams & Wilkins, Baltimore 1974).
45   Mendelson, J.H.; Mello, N.K.; Ellingboe, J.: Effects of alcohol on pituitary-gonadal hormones, sexual function, and aggression in human males; in Lipton, Lipton, DiMascio, Killam, Psychopharmacology. A generation of progress, pp. 1677–1692 (Raven Press, New York 1978).
46   Mendelson, J.H.; Mello, N.K.; Ellingboe, J.: Effects of acute alcohol intake on pituitary-gonadal hormones in normal human males. J. Pharmac. exp. Ther. *202:* 676–682 (1977).
47   Merry, J.: Endocrine response to ethanol. Int. J. ment. Hlth *5:* 16–28 (1976).
48   Mezey, E.: Ethanol metabolism and ethanol drug interactions. Biochem. Pharmac. *25:* 869–875 (1976).

49 Neims, A.H.; Bailey, J.; Aldridge, A.: Disposition of caffeine during and after pregnancy. Clin. Res. *27:* 236A (1979).
50 O'Malley, K.; Stevenson, I.H.; Crooks, J.: Impairment of human drug metabolism by oral contraceptive steroids. Clin. Pharmacol. Ther. *13:* 552–557 (1972).
51 Patkai, P.; Johannson, G.; Post, B.: Mood, alertness and sympathetic-adrenal medullary activity during the menstrual cycle. Psychosom. Med. *36:* 503–512 (1974).
52 Patwardhan, R.V.; Desmond, P.V.; Johnson, R.F.; Schenker, S.: Impaired elimination of caffeine by oral contraceptive steroids. J. Lab. clin. Med. *95:* 603–608 (1980).
53 Pincus, I.J.; Rakoff, A.E.; Cohn, E.M.; Tuman, H.J.: Hormonal studies in patients with chronic liver disease. Gastroenterology *19:* 735–754 (1951).
54 Ratnoff, O.D.; Patek, A.J.: The natural history of Laennec's cirrhosis of the liver. Medicine *21:* 207–265 (1942).
55 Redgrove, J.A.: Menstrual cycles; in Colquhoun, Biological rhythms and human performance (Academic Press, New York 1971).
56 Rubin, E.; Lieber, C.S.; Altman, K.; Gordon, G.G.; Southren, A.L.: Prolonged ethanol consumption increases testosterone metabolism in the liver. Science *191:* 563–564 (1976).
57 Sawin, C.T.: The hormones (endocrine physiology) (Little, Brown, Boston 1969).
58 Sommer, B.: The effect of menstruation on cognitive and perceptual-motor behavior. A review. Psychosom. Med. *35:* 515–534 (1973).
59 Stokes, P.E.: Alcohol-endocrine relationships; in Kissin, Begleiter, The biology of alcoholism, vol. I, pp. 397–436 (Plenum Press, New York 1971).
60 Tephly, T.R.; Mannering, G.J.: Inhibition of microsomal drug metabolism by steroid hormones. Pharmacologist *6:* 186 (1964).
61 Tephly, T.R.; Mannering, G.J.: Inhibition of drug metabolism, V. Inhibition of drug metabolism by steroids. Molec. Pharmacol. *4:* 10 (1968).
62 Van Thiel, D.H.: Alcohol and its effect of endocrine functioning. Alcohol. clin. exp. Res. *4:* 44–49 (1980).
63 Van Thiel, D.H.; Lester, R.: Sex and alcohol. New Engl. J. Med. *291:* 251–253 (1974).
64 Van Thiel, D.H.; Gaveler, J.S.; Lester, R.; Goodman, D.: Alcohol-induces testicular atrophy. An experimental model for hypogonadism occurring in chronic alcoholic men. Gastroenterology *69:* 326–332 (1975).
65 Van Thiel, D.H.; Gaveler, J.; Lester, R.: Ethanol inhibition of vitamin A metabolism in the testes. Possible mechanism for sterility in alcoholics. Science *186:* 941 to 942 (1974).
66 Van Thiel, D.H.; Sherins, R.J.; Lester, R.: Mechanism of hypogonadism in alcoholic liver disease. Gastroenterology *65:* 574 (1973).
67 Vierling, J.S.; Rock, J.: Variations in olfactory sensitivity to exaltalide during the menstrual cycle. J. appl. Physiol. *22:* 311–375 (1967).
68 Wiener, J.; Elmadjian, F.: Excretion of epinephrine and norepinephrine in premenstrual tension. Fed. Proc. *21:* 184 (1962).
69 Wilsnack, S.C.: Sex role identity in female alcoholism. J. abnorm. Psychol. *82:* 253 to 261 (1973).
70 Wineman, E.W.: Autonomic balance changes during the menstrual cycle. Psychophysiology *8:* 1–6 (1971).

71 Yoder, P.S.: Effects upon the palmar sweat index of hormonal changes related to the menstrual cycle. Nurs. Res. *19:* 165–168 (1970).
72 Zeiner, A.R.: Changes in autonomic responsivity related to the menstrual cycle. 14th Ann. Meet. Soc. Psychophysiological Research, Salt Lake City 1974.
73 Zeiner, A.R.; Davison, M.; Lovallo, W.: Autonomic responsivity as a function of the menstrual cycle. Southwestern Psychological Association Meeting, Dallas 1973.
74 Zeiner, A.R.; Farris, J.J.: Male-female and birth control pill effects on ethanol pharmacokinetics in American Indians. Alcohol. clin. exp. Res. *3:* 202 (1979).
75 Zeiner, A.R.; Kegg, P.S.: Menstrual cycle and oral contraceptive effects on alcohol pharmacokinetics in Caucasian females. Alcohol. clin. exp. Res. *4:* 233 (1980).

Dr. A.R. Zeiner, Department of Psychiatry and Behavioral Sciences,
University Hospital, Room 400E, PO Box 26901, Oklahoma City, OK 73190 (USA)

Marihuana and Alcohol: Perinatal Effects on Development of Male Reproductive Functions in Mice

S. Dalterio, A. Bartke, K. Blum, C. Sweeney

Departments of Pharmacology, Obstetrics and Gynecology and Anatomy, University of Texas, Health Science Center at San Antonio, San Antonio, Tex., USA

Introduction

Patterns of drug use among a broad spectrum of adolescents and younger adults include recreational exposure to caffeine, nicotine and alcohol, as well as access to certain illegal substances, most prominent of which is marihuana. Indeed, it has been estimated that approximately 20–30% of these regular users combine alcohol and marihuana [36]. The acute depressant effects of a combination of marihuana and alcohol are greater than those produced by either drug alone [45]. Synergistic effects of these two agents have also been observed in the production of anticonvulsant activity [25], impairment of motor performance [33], and depressant effects on heart rate and body temperature [39]. Other studies have dealt with pharmacokinetics [46], toxicity [50], and cross-tolerance of these two agents [48, 49].

However, little work is available concerning the reproductive consequences of the alcohol-marihuana combination. Several reports have indicated that, used alone, marihuana or alcohol can produce similar effects on reproductive processes in both man and laboratory animals. Thus, both marihuana and alcohol are capable of decreasing plasma testosterone concentrations, suppressing gonadotropin release, impairing spermatogenesis and interfering with libido and potency [14, 35, 41].

The mechanisms by which these alterations are produced remain unclear. Clinical studies in the human have often implicated poor nutritional habits together with alcohol-induced liver disease in the etiology of the loss of reproductive capacity in the chronic alcoholic [29, 53]. However, there have been reports that plasma testosterone levels are reduced during a 'hangover' in casual drinkers [56] and suppressed gonadotropins have also been observed after acute alcohol exposure in man [37]. It has, therefore, been proposed that alcohol, or a metabolite such as acetaldehyde, may exert direct inhibitory effects at all levels of the hypothalamo-pituitary-gonadal axis [15]. Marihuana has been shown to suppress the pituitary release of luteinizing hormone (LH) and follicle-stimulating hormone (FSH) in several species [9, 17] and reduced pituitary-responsiveness to hypothalamic luteinizing hormone releasing hormone (LHRH), as well as symptoms of depressed hypothalamic function, have been reported after cannabinoid administration [2, 9].

Psychoactive and nonpsychoactive components in marihuana decrease testosterone production in decapsulated testes [16] and isolated Leydig cells in mice, possibly as a result of cholesterol esterase inhibition [11]. Various cannabinoids have also been found to inhibit protein, lipid and nucleic acid synthesis in rat testicular slices [31]. Alcohol was found to have little effect in these in vitro systems [16], however, acetaldehyde was in fact inhibitory in testes from mice [3] and rats [15]. It has also been suggested that ethanol inhibition of testicular vitamin A metabolism may cause infertility in alcoholics [52].

One of the most recent areas of investigation concerning effects of abused substances on reproductive processes followed the description of the 'fetal alcohol syndrome' (FAS). This condition includes embryonic, fetal or neonatal death, intrauterine and postnatal growth retardation, congenital malformations and behavioral effects in infants whose mothers used alcohol regularly [32].

Alcohol rapidly diffuses across the placental barrier into the fetus, resulting in concentrations comparable to those found in maternal circulation [30, 47]. The development of laboratory animal models for FAS has resulted in evidence that prenatal alcohol exposure can produce alterations in brain amines [1, 13, 24, 40], which may be related to impaired brain growth, as well as the mental impairment often characteristic of FAS children. In mice, early alcohol exposure affects susceptibility to audiogenic seizures, aggressive and open field behaviors, and hepatic metabolizing enzymes [51, 55].

In rats prenatal alcohol exposure prolonged gestation, and decreased anogenital distance in pups of both sexes at birth, although adult male sexual behavior, plasma testosterone and weights of sex accessory glands were not affected [12].

Several studies have been directed toward determination of the teratogenic potential of marihuana, particularly the Δ^9-tetrahydrocannabinol (THC) component. Few studies have observed congenital abnormalities unless the doses employed were extremely high. Several studies present evidence of reduced litter size and birth weight after exposure to moderate amounts of cannabinoids, suggesting that these substances can be embryotoxic [26]. Rats exposed to cannabis smoke delivered pups who later displayed delayed eye opening and incisor eruption and reduced weight gain [27]. Sex differences in later learning ability have been observed in rats exposed to marihuana extract prenatally [27, 28]. In addition, male Rhesus monkeys prenatally exposed to THC showed behavioral deficits, though females exhibited higher mortality [43]. Therefore, it is possible that hormonal status may moderate the effects of cannabinoids on the central nervous system. In earlier studies we have shown that maternal exposure to THC or to cannabinol (CBN), a nonpsychoactive cannabinoid, results in long-term alterations in body weight regulation and pituitary-gonadal function, as well as in adult sexual behavior in their male offspring [18, 19]. The mechanism of cannabinoid action on male sexual differentiation is not well understood. It is possible that THC or CBN may interfere with LH release by the fetal pituitary, as they do in adult mice [17, 19], thereby suppressing testicular steroid production. In preliminary studies, exposure to THC or CBN during mid-gestation reduced fetal testosterone concentration in mice [20]. Interference with testicular androgen production during critical periods of sexual differentiation has been shown to affect development of the male reproductive structures, fertility, steroid uptake in various brain regions, target organ sensitivity to postpubertal hormone concentrations and sex-typical behavioral responses [6]. Therefore, it is possible that early exposure to teratogenic agents may alter physiologic processes in a way not apparent at birth or until challenged by later developmental events, such as puberty or pregnancy, as has been observed in human females subsequent to in utero exposure to diethylstilbestrol [34].

These studies were conducted to investigate the consequences of combined maternal exposure to THC and alcohol (ETOH) on the development of reproductive functions in their male offspring.

Experimental Design

Animals

Random-bred mice were housed on a 14 h light:10 h dark lighting schedule and provided Wayne Breeder Blox and tap water ad libitum. Castrations were performed through a mid-ventral incision under ether anesthesia.

Drugs

Ethanol (40% w/v; 0.33 g/kg body weight) and ETOH in combination with THC (50 mg/kg body weight) were administered by oral feeding in 50 µl volume after appropriate aqueous dilution. Mice receiving THC alone received the same dose in sesame oil. Two groups of control animals were given either 50 µl of sesame oil or tap water.

Radioimmunoassay

Blood was obtained by cardiac puncture under ether anesthesia and plasma was stored frozen for radioimmunoassay of testosterone [7], without chromatography [4], and LH and FSH using Niswender's anti-ovine LH serum and NIAMDD rat LH and FSH kits which have been validated for measuring mouse gonadotropins [8], and as described previously [19].

Procedures

Adult primiparous female mice were housed with an adult male and checked daily for the presence of a vaginal plug (considered day 1 of pregnancy). Females were treated with ETOH, THC, ETOH plus THC or vehicles (water or sesame oil), beginning approximately 1 day prior to parturition. The second treatment was administered on the day of parturition and the treatment was continued daily for 6 days postpartum (a total of 7 daily treatments). At birth, litters were culled to 6 male pups and the young were weaned at 21 days of age and housed 3 per cage until adulthood (60–70 days of age). A group of males from each group were castrated and testes were weighed, decapsulated and incubated for 4 h in buffer containing human chorionic gonadotropin (hCG) (12.5 mIU/ml) as described previously [16]. The remaining animals were bled and sacrificed by cervical dislocation and tissue weights were recorded. Castrated animals were bled 3 weeks later for LH and FSH determinations.

Statistics

Data were analyzed using analysis of variance (Anova) with Duncan's multiple range test for paired comparison [54] or Wilcoxon's rank sum test [44].

In each experiment, there were no differences between control groups given different vehicles with respect to any parameter measured, therefore, the results obtained in these groups were combined for statistical analysis and graphic presentation.

Results

In adult male mice perinatally exposed to ETOH, THC, or ETOH plus THC, testes weights were significantly reduced ($p<0.05$), but

Table I. Organ weights in adult male mice perinatally exposed to ETOH, THC or ETOH plus THC; means ± SE [from ref. 23]

	n	Body weight g	Testes mg	Seminal vesicles (full) mg	Adrenal mg	Kidneys mg
Control[1]	16	34 ± 2	324 ± 12[2]	298 ± 20[2]	4.8 ± 0.15	626 ± 13[2]
ETOH	12	32 ± 1	255 ± 18[3,4]	293 ± 13[2]	4.4 ± 0.40	463 ± 11[3]
THC	12	34 ± 2	253 ± 6[4]	236 ± 35[2]	5.0 ± 0.3	533 ± 8[4]
ETOH + THC	12	33 ± 1	284 ± 8[3]	284 ± 7[2]	4.6 ± 0.23	637 ± 18[2]

[1] The vehicle control groups did not differ from one another and were therefore combined for presentation and statistical analysis.
[2,3,4] Values with the same superscript not significantly different ($p<0.05$) by one-way Anova and Duncan's test [54].

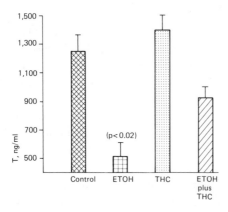

Fig. 1. Testosterone production in vitro in the presence of hCG (12.5 mIU/ml) by decapsulated testes from mice perinatally exposed to THC, ETOH or ETOH plus THC. Means ± SE (n = 6) [from ref. 23].

seminal vesicle weights were reduced only in males perinatally exposed to THC (table I). Kidney weights were reduced in adult males perinatally exposed to THC or ETOH, but not in those exposed to ETOH plus THC. Adrenal weights and body weight were not affected by any perinatal drug exposure.

The accumulation of testosterone in respone to hCG stimulation in vitro was reduced ($p<0.05$) in decapsulated testes obtained from mice perinatally exposed to ETOH (fig. 1). However, in vitro testosterone

Table II. Plasma levels of testosterone (T), LH and FSH from adult male mice perinatally exposed to ETOH, THC or ETOH plus THC, and hCG (12.5 mIU/ml) stimulated T production after 4 h incubation of decapsulated testes from males in these treatment groups, Means ± SE (n) [from ref. 23].

	Plasma T (ng/ml)	Plasma LH	Plasma FSH
Control	8.1 ± 2.4 (7)	50 ± 8[2] (9)	1,319 ± 91[2] (7)
Control, castrated[1]	–	223 ± 28[3] (8)	2,243 ± 171[3] (8)
ETOH	2.4 ± 1.7 (6)[6]	49 ± 10[2] (7)	1,498 ± 142[2] (8)
ETOH, castrated[1]	–	277 ± 23[3,4] (8)	1,944 ± 99[4] (8)
THC	2.6 ± 0.8 (8)[6]	54 ± 3[2] (10)	1,386 ± 66[2] (7)
THC, castrated[1]	–	304 ± 36[4] (10)	3,026 ± 218[5] (10)
ETOH + THC	2.8 ± 0.8 (8)[6]	28 ± 7[2] (8)	1,602 ± 72[3] (7)
ETOH + THC, castrated[1]	–	276 ± 37[3,4] (10)	2,280 ± 205[4] (8)

[1] 3 weeks postcastration.
[2,3,4,5] Values with the same superscript not significantly different by Anova, $p < 0.05$.
[6] $p < 0.05$ using Wilcoxon's sum rank test [44].

production by testes of THC-treated mice was slightly elevated, while testosterone secretion by testes from mice perinatally exposed to ETOH plus THC was intermediate between the levels obtained in animals receiving ETOH or THC separately.

Exposure to any of these drugs, alone or in combination, resulted in a significant reduction in plasma testosterone levels. In addition, plasma LH levels were reduced and those of FSH increased in the animals exposed to ETOH plus THC (table II). 3 weeks postcastration, peripheral levels of both LH and FSH were elevated in THC-exposed as compared to castrated control mice. The gonadotropic response to castration was not influenced by perinatal exposure to ETOH or ETOH plus THC.

Conclusions

Perinatal exposure to ETOH or THC appears to exert complex interactive effects on the development of male reproductive functions in mice.

Treatment with either ETOH or THC during the perinatal period resulted in reduced testicular and renal weights in adulthood, while in combination, they were much less effective, in that no alteration or a lesser effect, respectively, was produced. Exposure to THC, but not in combination with ETOH, reduced weights of the seminal vesicles. In addition, perinatal exposure to THC resulted in elevated plasma FSH levels in adult mice 3 weeks after castration, whereas the response to castration in males exposed to both ETOH plus THC was comparable to controls. Although plasma testosterone levels were reduced by perinatal exposure to ETOH or THC, alone or in combination, the production of testosterone by testes in vitro was differentially affected by these treatments. These effects may be due to ETOH and THC affecting different processes involved in testicular responsiveness to gonadotropic stimulation.

Perinatal exposure to THC has several consequences for the development of reproductive functions in mice [19, 22, 23]. In both our earlier studies and the present experiments, we observed decreased testes and seminal vesicle weights in adult males after perinatal THC exposure. In the present study, plasma FSH levels were also elevated 3 weeks postcastration in THC-exposed males. Previously, we observed a significant elevation in plasma LH, but no change in plasma testosterone, while in the present experiments plasma LH levels were normal, and plasma testosterone levels were reduced. Both situations are compatible with the suggestion that THC decreases testicular responsiveness to gonadotropin stimulation and interferes with pituitary-gonadal feedback [17–21].

The mechanism of these cannabinoid actions on the development of male reproductive functions remains to be elucidated. We have recently observed a reduction in the in vitro responsivity of the vas deferens to norepinephrine and enkephalin in adult male mice who had been exposed to THC during the perinatal period [22]. In preliminary studies (fig. 2), we have not found evidence to indicate that perinatal ETOH alone affects the response of the vas to enkephalin. In the ETOH combination with THC, the effect appear to be explicable on the basis of the presence of THC alone, as reported earlier [22].

It is possible that the effects of ETOH or THC may be antagonistic on some parameters, since a decrease in kidney weight was observed with exposure to either ETOH or THC, but not both. We have observed pathological changes, e.g., substantial size differences, hydronephrosis, degenerative changes etc., in a few kidneys obtained from mice exposed perinatally or during sexual maturation to ETOH [23]. Renal anomalies have been observed after prenatal ETOH treatment in both male and female mice [10]. The alterations in kidney growth and function may be

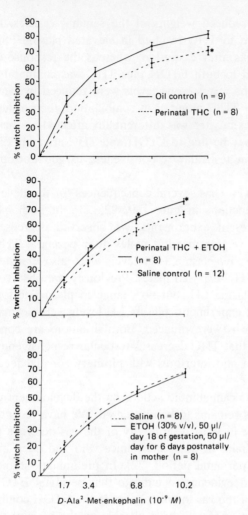

Fig. 2. The response of vasa deferentia obtained from adult male mice to D-Ala-^{2}-Met-enkephalin in vitro. These mice were exposed to THC (top) [from ref. 22], ETOH (middle) or their combination (bottom) via maternal exposure (means ± SE). *Significantly different from controls ($p < 0.02$) by Student's t test.

due to the decrease in plasma testosterone in ETOH and THC-treated mice [4, 17]. Kidney weight exhibits marked sexual dimorphism in the mouse, and it has been demonstrated that testicular androgens are responsible for the larger kidney weight characteristic of the adult male [5]. Androgen levels have also been reported to influence ETOH preference [38] and ETOH exposure can increase the metabolism of testosterone [42].

The ability of ETOH and THC each to influence the actions of the other may be related to reported cross-tolerance. The mechanism of this cross-tolerance may occur at the cellular level with local changes in membrane permeability, or alterations in absorption, distribution and metabolism being involved [25, 47, 49]. Cellular actions may be relevant to the possible direct effect of ETOH [15] and THC [17] on the testis, which may be exacerbated when these substances are given in combination.

The present findings indicate that perinatal exposure to ETOH or THC, alone or in combination, during the perinatal period can influence the development of renal and pituitary-gonadal function in male mice.

Summary

Marihuana and alcohol have been reported to exert a wide range of effects on reproductive functions in man and laboratory animals. Decreased testosterone levels, suppression of pituitary gonadotropin release, impaired fertility and sexual dysfunction have been observed in adult males exposed to either of these substances.

The present studies examine the effects of perinatal exposure to ETOH and THC alone or in combination on the development of reproductive functions in mice. Testes weights and plasma testosterone levels were reduced in adult males perinatally exposed to these substances, alone or in combination. Seminal vesicle weights were reduced only in the THC-exposed mice, while kidney weights were decreased by either ETOH or THC exposure. In ETOH plus THC-exposed animals, plasma LH was reduced while FSH levels were increased. In response to castration, gonadotropin levels were elevated in mice perinatally exposed to THC. In vitro responsiveness to gonadotropin stimulation, as indicated by testosterone production by decapsulated testes, was significantly inhibited in tissue obtained from ETOH-exposed males.

Although no simple interactions were observed, it is evident that THC and ETOH, either alone or in combination, can affect the development of male reproductive functions in mice.

References

1 Abel, E.: Effects of ethanol exposure during different gestation weeks of pregnancy on maternal weight gain and intrauterine growth retardation in the rat. Neurobehav. Toxicol. *1:* 145–151 (1979).

2 Asch, R.H.; Smith, C.G.; Silerkhodr, T.M.; Pauerstein, C.J.: Acute decreases in serum prolactin concentrations caused by Δ^9-tetrahydrocannabinol in nonhuman primates. Fert. Steril. *32:* 571–575 (1979).

3   Badr, F.M.; Bartke, A.; Dalterio, S.; Bulger, W.: Suppression of testosterone production by ethyl alcohol. Possible mode of action. Steroids *30:* 647–655 (1972).
4   Badr, F.M.; Bartke, A.: Effect of ethyl alcohol on plasma testosterone level in mice. Steriods *23:* 921–928 (1974).
5   Bardin, C.W.; Brown, T.R.; Mills, N.C.; Gupta, C.; Bullock, L.P.: The regulation of β-glucuronidase gene by androgens and progestins. Biol. Reprod. *18:* 74–83 (1978).
6   Barraclough, C.A.: Hormones and development. Modification in the CNS regulation of reproduction after exposure of prepubertal rats to steroid hormones. Recent Prog. Horm. Res. *22:* 503–539 (1966).
7   Bartke, A.; Steele, R.E.; Musto, N.; Caldwell, B.V.: Fluctuations in plasma testosterone levels in adult male rats and mice. Endocrinology *92:* 1223–1228 (1973).
8   Beamer, W.G.; Murr, S.M.; Geschwind, I.I.: Radioimmunoassay of mouse luteinizing and follicle stimulating hormones. Endocrinology *90:* 823–826 (1972).
9   Bloch, E.; Thysen, B.; Morrill, G.A.; Gardner, E.; Fujimoto, G.: Effects of cannabinoids on reproduction and development. Vitam. Horm. *36:* 203–258 (1978).
10 Boggan, W.O.; Randall, C.L.; Beukelauer, M. de; Smith, R.: Renal anomalies in mice prenatally exposed to ethanol. Res. Commun. chem. Pathol. Pharmacol. *23:* 127–142 (1979).
11 Burstein, S.; Hunter, S.A.; Shoupe, T.S.: Inhibition of cholesterol esterases by Δ^1-tetrahydrocannabinol. Life Sci. *23:* 972–979 (1978).
12 Chen, J.J.; Smith, E.R.: Effects of perinatal alcohol on sexual differentiation and open-field behavior in rats. Horm. Behav. *13:* 219–231 (1979).
13 Chernoff, G.F.: The fetal alcohol syndrome in mice: an animal model. Teratology *15:* 223–229 (1977).
14 Cicero, T.J.; Badger, T.M.: A comparative analysis of the effects of narcotics, alcohol and the barbiturates on the hypothalamic-pituitary-gonadal axis; in Gross, Alcohol intoxication and withdrawal. IIIb. Studies in alcohol dependence, pp. 95–115 (Plenum, New York 1977).
15 Cicero, T.J.; Meyer, E.R.; Bell, R.D.: Effects of ethanol on the hypothalamic-pituitary-luteinizing hormone axis and testicular steroidogenesis. J. Pharmac. exp. Ther. *208:* 210–215 (1979).
16 Dalterio, S.; Bartke, A.; Burstein, S.: Cannabinoids inhibit testosterone secretion by mouse testes in vitro. Science *196:* 1472–1473 (1977).
17 Dalterio, S.; Bartke, A.; Roberson, C.; Watson, D.; Burstein, S.: Direct and pituitary-mediated effects of Δ^9-THC and cannabinol on the testis. Pharmacol. Biochem. Behav. *8:* 673–678 (1978).
18 Dalterio, S.; Bartke, A.: Perinatal exposure to cannabinoids alters male reproductive function in mice. Science *205:* 1420–1422 (1979).
19 Dalterio, S.L.: Perinatal or adult exposure to cannabinoids alters male reproductive functions in mice. Pharmacol. Biochem. Behav. *12:* 143–153 (1980).
20 Dalterio, S.; Bartke, A.: Fetal testosterone in mice: effect of gestational age and cannabinoid exposure. 61st Ann. Meet. Endocrine Soc. No. 23, p. 78 (1979).
21 Dalterio, S.: Marihuana and male reproduction. Substance Alcohol Actions/Misuse *2:* 1–14 (1981).

22 Dalterio, S.; Blum, K.; Delallo, L.; Sweeney, C.; Briggs, A.; Bartke, A.: Perinatal exposure to Δ^9-THC in mice: Altered enkephalin and norepinephrine sensitivity in vas deferens. Substance Alcohol Actions/Misuse *1:* 467–468 (1980).

23 Dalterio, S.; Bartke, A.; Sweeney, C.: Interactive effects of ethanol and Δ^9-tetrahydrocannabinol on endocrine functions in male mice. J. Androl. *2:* 87–93 (1980).

24 Diaz, J.; Samson, H.H.: Impaired brain growth in neonatal rats exposed to ethanol. Science *208:* 751–753 (1980).

25 Esplin, B.; Capek, R.: Quantitative characterization of THC and ethanol interaction. Res. Commun. chem. Pathol. Pharmacol. *15:* 199–202 (1976).

26 Fleischman, R.W.; Hayden, D.W.; Rosenkrantz, H.; Braude, M.C.: Teratologic evaluation of Δ^9-THC in mice, including a review of the literature. Teratology *12:* 47–50 (1975).

27 Fried, P.A.: Short and long term effects of prenatal cannabis inhalation upon rat offspring. Psychopharmacology *50:* 285–291 (1976).

28 Gianutsos, G.; Abbatiello, E.R.: The effect of pre-natal Canabis Sativa on maze learning in the rat. Psychopharmacology, Berlin *27:* 117–122 (1974).

29 Gordon, G.G.; Southren, A.L.; Lieber, C.S.: The effects of alcoholic liver disease and alcohol ingestion on sex hormone levels. Alcoholism. Clin. exp. Res. *2:* 259–263 (1978).

30 Idanpaan-Heikkila, J.; Jouppila, P.; Akerblom, H.K.; Isaho, R.; Kauppila, E.; Koivisto, M.: Elimination and metabolic effects of ethanol in mother, fetus and newborn infant. Am. J. Obstet. Gynec. *112:* 387–393 (1972).

31 Jakubovic, A.; McGeer, P.L.: In vitro inhibitions of protein and nucleic acid synthesis in rat testicular tissue by cannabinoids; in Nahas, Marihuana: chemistry, biochemistry and cellular effects, pp. 223–241 (Springer, New York 1976).

32 Jones, K.L.; Smith, D.W.: Recognition of the fetal alcohol syndrome in early infancy. Lancet *ii:* 999–1001 (1973).

33 Kalant, H.; Blanc, A.E. le: Effect of acute and chronic pretreatment with Δ^1-tetrahydrocannabinol on motor impairment by ethanol in the rat. Can. J. Physiol. Pharmacol. *52:* 291–297 (1974).

34 Kaufman, R.H.; Binder, G.L.; Gray, P.M.; Adam, E.: Upper genital tract changes associated with exposure in utero to diethyl-stilbestrol. Am. J. Obstet. Gynec. *128:* 51–59 (1977).

35 Kolodny, R.C.; Masters, W.H.; Kolodner, R.M.; Toro, G.: Depression of plasma testosterone levels after chronic intensive marihuana use. New Engl. J. Med. *290:* 872–874 (1974).

36 Mello, N.K.; Mendelson, J.H.; Kuehnle, J.C.; Sellers, M.L.: Human polydrug use: marihuana and alcohol. J. Pharmac. exp. Ther. *207:* 922–935 (1978).

37 Mendelson, J.H.; Mello, N.K.; Ellingboe, J.: Effects of acute alcohol intake on pituitary-gonadal hormones in normal human males. J. Pharmac. exp. Ther. *202:* 676–682 (1977).

38 Messiha, F.S.: Androgens, antiandrogens and voluntary intake of ethanol by the male rat. Res. Commun. Subst. Abuse *1:* 1–8 (1980).

39 Pryor, G.T.; Larsen, F.F.; Carr, J.D.; Braude, M.C.: Interactions of Δ^9-tetrahydrocannabinol, ethanol, and chlordiazepoxide. Pharmacol. Biochem. Behav. *7:* 331–345 (1977).

40   Rawat, A.K.: Effects of maternal ethanol consumption on the fetal and neonatal cerebral neurotransmitters; in Lindros, Eriksson, The role of acetaldehyde in the action of ethanol, pp. 237 (Finnish Foundation for Alcohol studies, Helsinki; distributed by Rutgers University Center of Alcohol Studies, New Brunswick 1975).
41   Rubin, H.B.; Henson, D.E.: Effects of alcohol on male sexual responding. Psychopharmacol. Bull. *47:* 123–124 (1976).
42   Rubin, E.; Leber, C.S.; Altman, K.: Prolonged ethanol consumption increases testosterone metabolism in the liver. Science *191:* 563–564 (1976).
43   Sassenrath, E.N.; Chapman, L.: Tetrahydrocannabinol-induced manifestation of the 'marihuana syndrome' in group-living Macaques. Fed. Proc. *34:* 1666–1670 (1975).
44   Siegel, S.: Nonparametric statistics for the behavioral sciences (McGraw-Hill, New York 1956).
45   Siemens, A.; Khanna, J.M.: Acute metabolic interaction between ethanol and cannabis. Alcoholism. Clin. exp. Res. *1:* 343–348 (1977).
46   Siemens, A.J.; Doyle, O.L.: Cross tolerance between Δ^9-tetrahydrocannabinol and ethanol: the role of drug disposition. Pharmacol. Biochem. Behav. *10:* 49–56 (1979).
47   Sippell, H.W.; Kesaniemi, Y.A.: Placental and foetal metabolism of acetaldehyde in rat. II. Studies on metabolism of acetaldehyde in the isolated placenta and foetus. Acta pharmac. tox. *37:* 49–55 (1975).
48   Sofia, R.D.; Knobloch, L.C.: The interaction of Δ^9-tetrahydrocannabinol pretreatment with various sedative hypnotic drugs. Psychopharmacologia *30:* 185–194 (1973).
49   Sprague, G.L.; Craigmill, A.L.: Ethanol and delta-9-tetrahydrocannabinol: mechanism for cross tolerance in mice. Pharmacol. Biochem. Behav. *5:* 409–415 (1976).
50   Sulkowski, A.; Vachon, L.: Side effects of simultaneous alcohol and marihuana use. Am. J. Psychiat. *134:* 691–692 (1977).
51   Sze, P.Y.; Yanai, J.; Ginsburg, B.E.: Effects of early ethanol input on the activities of ethanol metabolizing enzymes in mice. Biochem. Pharmac. *25:* 215–217 (1976).
52   Thiel, D.H. van; Gavaler, J.; Lester, R.: Ethanol inhibition of vitamin A metabolism in the testes: possible mechanism for sterility in alcoholics. Science *186:* 941–942 (1974).
53   Thiel, D.H. van; Gavaler, J.; Lester, R.; Goodman, M.D.: Alcohol-induced testicular atrophy. Gastroenterology *69:* 326–332 (1975).
54   Winer, B.J.: Statistical principles in experimental design (McGraw-Hill, New York 1962).
55   Yanai, J.; Ginsburg, B.E.: Long term reduction of male agonistic behavior in mice following early exposure to ethanol. Psychopharmacology *52:* 31–34 (1977).
56   Ylikahri, R.; Huttunen, M.; Harkonen, M.; Suederling, V.; Onikki, S.; Karonen, S.L.; Adlercreutz, H.: Low plasma testosterone values in men during hangover. J. Steroid Biochem. *5:* 655–658 (1974).

Dr. S. Dalterio, Departments of Pharmacology, Obstetrics and Gynecology,
The University of Texas Health Science Center at San Antonio,
7703 Floyd Curl Drive, San Antonio, TX 78284 (USA)

Subcellular Fractionation of Alcohol and Aldehyde Dehydrogenase in the Rat Testicles[1]

F.S. Messiha

Division of Toxicology, Department of Pathology, and Psychopharmacology Laboratory, Department of Psychiatry, Texas Tech University Health Sciences Center, School of Medicine, Lubbock, Tex., USA

Introduction

The adverse effects of ethyl alcohol (ET) on the endocrine function continue to receive increasing attention over the years. The primary interest has been focused on the effect of ET on hypothalamic-pituitary-adrenocortical axis, hypothalamic-pituitary-gonadal function and on the hypothalamic-pituitary-thyroid relationship [1, 3, 11, 12]. One approach for studying the toxicity of ET on the gonads may reside in the evaluation of its metabolic pathway in the target organs. This requires knowledge of biotransformation routes and properties of the enzymes involved. The metabolic detoxification of ET proceeds by its oxidation to acetaldehyde by the action of alcohol dehydrogenase (ADH; EC 1.1.1.1) and the subsequent oxidation of the acetaldehyde formed to acetate by aldehyde dehydrogenase (ALDH; EC 1.2.1.3) prior to the condensation process to form acetyl Co-A. The latter can be metabolized by a number of routes, i.e. oxidation through the citric acid cycle, utilization in biosynthesis of fatty acids and in cholesterol formation. Both ADH and ALDH are widely distributed in the organism and the rat testis contains ADH, as demonstrated histochemically [2] and biochemically [10] in whole homogenates, and also ALDH [6–9].

---

[1] Supported in part by a grant from TTUHSC Biomedical Research Institute.

The present study reports on the intercellular distribution of ADH and ALDH in the rat testicles. The kinetic properties of the testicular enzymes is also reported in conjunction with their in vitro response to pharmacological interventions.

Material and Methods

Male Sprague-Dawley rats (Holtzman Farm Co., Madison, Wisc.), 70–90 days old, were used throughout the experiments. Animals had access to water and purina pellet food and were housed in a room with alternating 12 h dark and light cycles.

Fractionation of testicular and epididymal tissues: The differential centrifugation technique was used to obtain the various intercellular components of the rat testis and epididymis as outlined in figure 1. Animals were sacrificed by decapitation and the testis and the epididymis were quickly removed, rinsed with ice-cold $0.1\,M$ KCl buffer pH 6.8, blotted dry and weighed. A pool of 5–7 individual pairs of testis or epididymis were homogenized in the KCl buffer to obtain 10% (w/v) homogenate by Potter-Elvehjem glass homogenizer. Aliquots of the homogenates obtained were treated with $0.1\,M$ sodium desoxycholate. The homogenate was subjected to differential centrifugation procedure as outlined in figure 1. This consisted of centrifugation of the total homogenate at $200\,g$ for 25 min and washing the resulting pellet, by its resuspension twice in the ice-cold KCl buffer and recentrifugation at $200\,g$ for 25 min, prior to its solubilization with $0.1\,M$ sodium desoxycholate to obtain the nuclei fraction (NC). The $200\,g$ supernatant fluid was then centrifuged at $500\,g$ for 25 min and the pellet was similarly treated by successive washing twice with the KCl buffer and subsequent centrifugation at $500\,g$ for 25 min each. The washed pellet was solubilized with the sodium desoxycholate solution and constituted the mitochondrial fraction (MT). The $500\,g$ supernatant was centrifuged for 90 min at $22,000\,g$ to obtain the initial cytosolic fraction (CT) and the pellet was discarded. An aliquot of the CT fraction was centrifuged for 60 min at $100,000\,g$. The resulting cytoplasmic fraction (CT) was saved and the pellet was resuspended in the KCl buffer and recentrifuged for a further 60 min at $100,000\,g$. The washed pellet was suspended in a small volume of KCl buffer and served as the microsomal fraction (MC). All fractionations were performed at $2-4\,°C$ and the enzymatic assays were made in the cellular components obtained within 4–6 h of their isolation, unless otherwise specified. The volumes of the fractions were measured and their protein content was determined by the biuret procedure. The enzymatic activity is expressed as specific activity (nmol/min/mg protein) and was measured in a double beam recording spectrophotometer by following the changes in optical density of the cofactor as a function of time at $30\,°C$.

Aliquots from NC, MT, CT and MC fractions were measured for their contents of NADP- and NADPH-dependent ALDH. The NADP-dependent ALDH was assayed using $0.1\,M$ phosphate buffer pH 7.0 while that of NADP-linked ALDH was measured at pH 9.8 using $0.1\,M$ sodium pyrophosphate buffer. The assay's mixture consisted of excess of NADP or NADPH in the absence (assay's blank) of or in the presence of acetaldehyde as substrate. The NAD-dependent ALDH was also measured as described previously [8]. Pyrazol, $0.15\,M$, was present in all reaction mixture used for determinations of ALDH.

Testicular Fractionation

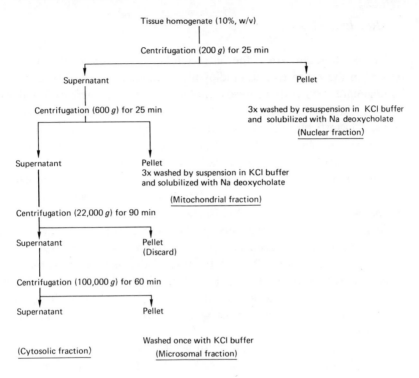

Fig. 1. Fractionation procedure.

Aliquots from the MC fraction were also assayed for the microsomal ethanol oxidizing system using NADH as cofactor and 0.1 M phosphate buffer pH 7.4 [4]. The ADH was also measured in NC and CT fractions of the rat testis using NAD as cofactor.

Experiments related to determinations of maximal enzymatic reaction requirements evaluated the effects of protein concentration (mg protein/ml assay mixture) and the assay's temperature on the velocity of the enzymatic reaction. The effects of various buffers and their pH ranges, i.e. sodium pyrophosphate buffer, Tris-HCl buffer and Soerensen phosphate buffer, on the reaction was also studied. In addition, the stability of enzymatic activity in the various cellular preparations was studied as a function of storage at $-20\,°C$.

The kinetic properties, i.e. K_m and V_{max}, of ADH and ALDH were determined in testicular fractions with demonstrable activity using the *Lineweaver and Burk* [5] methods. The in vitro effects of some selected agents on specific activity of testicular ADH was also evaluated. These were the O-methylated amine and acidic metabolites of the catecholamines, histamine diphosphate, the histamine antagonist, cimetidine and the CNS stimulants, d-amphetamine and aminophylline.

The results are expressed as means specific activity (nmol/min/mg protein) ± SE. Statistical analyses were performed, where applicable, using Student's t test for independent means.

Results

Table I summarizes the intercellular distribution of ADH and ALDH in the rat testis and epididymis. Solubilization of testicular tissue homogenate with sodium desoxycholate resulted in increased activities of both testicular ADH and ALDH but not of epididymal ALDH. This increase in testicular ADH and ALDH was approximately 1.2- and 1.8-fold from nonsolubilized homogenate, respectively. There was approximately 2.6-fold greater specific activity of ALDH than ADH in whole testicular homogenate. Testicular ALDH was present in all intercellular components fractionated compared to that of ADH which was measurable in the CT and in the NC fractions. The highest specific activity of testicular ALDH was determined in the NC and the CT fractions. Conversely, the lowest ALDH, 2.7 ± 0.1 nmol/min/mg protein, was present in the MC preparation. The MT fraction of the testis showed moderate specific activity between those of the CT and the MC fractions. In general, specific activity of testicular ALDH was greater than testicular ADH in the same cellular component where both were present, i.e. in the cytoplasma.

Table I also shows the localization of ALDH in the rat epididymis. The epididymis was devoid of measurable ADH activity compared to

Table I. Subcellular distribution of alcohol and aldehyde dehydrogenase activities in the adult rat testis and epididymis (nmol/min/mg protein)

Fraction	T-ADH	T-ALDH	EP-ALDH
Homogenate 10% (w/v)	1.3 ± 0.2 (4)	3.4 ± 0.2 (7)	3.3 ± 1.2 (4)
Nuclear	3.8 ± 0.7 (5)	6.2 ± 0.4 (9)	NM (5)
Mitochondrial	NM (5)	4.3 ± 0.3 (5)	NM (5)
Cytosolic			
(22,000 g supt.)	4.6 ± 0.7 (5)	6.6 ± 0.1 (5)	8.3 ± 0.8 (5)
(100,000 g supt.)	2.7 ± 1.3 (5)	7.1 ± 0.3 (5)	14.7 ± 1.1 (5)
Microsomal	NM (5)	2.7 ± 0.1 (5)	NM (4)

Rats were 90–110 days old and each single determination derives from a pool of 5 pairs of testis or epididymis. Values are means ± SE of specific activity of testicular alcohol dehydrogenase (T-ADH), aldehyde dehydrogenase (T-ALDH) and epididymal aldehyde dehydrogenase (EP-ALDH) for the number of determinations given between parentheses. Samples showing changes in optical density less than twice the blank are considered non-measurable (NM).

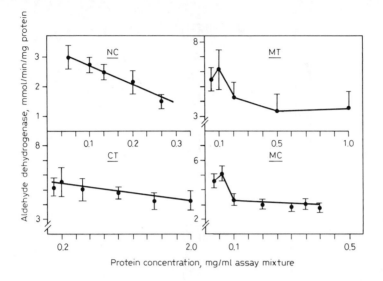

Fig. 2. The effect of protein concentration on the velocity of ALDH reaction in various testicular subcellular fraction. Values are means ± SE of specific activity of ALDH measured in the nuclear (NC), mitochondrial (MT), cytosolic (CT) and microsomal (MC) fractions of the rat testis. Each point represents the mean of 6–8 independent assays.

higher specific activity for ALDH. The latter was confined to the CT fraction and the ALDH activity assayed was greater than that of ALDH determined in the testicular cytoplasma.

Figure 2 shows the effect of protein concentration on the velocity of the ALDH reaction for the various intercellular components of the rat testicular fraction. The velocity of the reaction decreased as a function of increasing the protein concentrations in the assay mixture. A protein concentration exceeding 0.06 or 0.1 mg/ml of the reaction mixture tends to reduce microsomal and mitochondrial ALDH, respectively. This is compared with a decrease in cytosolic ALDH activity in concentration above 0.1 mg/ml of assay mixture (fig. 2, lower left panel).

Figure 3 shows the effect of the assay's temperature on the specific activity of testicular ALDH in the intercellular fractions indicated. In general, a temperature as low as 4 °C or a temperature as high as 50 °C decreased the measurable enzymatic activity compared to ALDH activity determined at 30 °C. This was evident in all preparations studied. Testicular ADH showed also a similar pattern to that of testicular, ALDH in response to the assay's temperature.

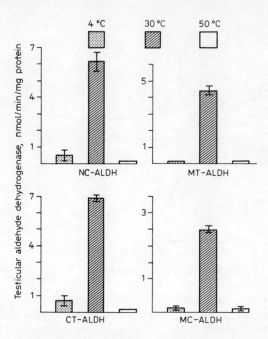

Fig. 3. The effect of assays temperature on the specific activity of ALDH in subcellular component of the rat testis. Values are means ± SE of specific activity of ALDH measured in the nuclear (NC), mitochondrial (MT), cytosolic (CT) and microsomal (MC) fractions of the rat testis. Each bar graph represents the mean ± SE of 5 independent determinations.

Figure 4 shows the stability of ALDH during storage at −20°C for the testicular fractions and the supernatant fluids isolated. Freshly prepared fractions showed higher ALDH activity than that assayed in the same fraction but after storage overnight at −20°C. Both cytoplasmic ALDH and 4,500 g supernatant were most stable under these conditions. The activity of testicular ADH was completely destroyed by freezing for 16 h at −20°C.

Figure 5 illustrates reciprocal plots of substrate concentration vs. reaction velocity of testicular ALDH in the rat NC, MT, CT and MC fractions. The V_{max} was lowest in the MC fraction, 3.4 mM, compared to the highest mean value of 5.9 mM measured in the NC fraction. The V_{max} for both the MT and CT component of the testis were of similar order and magnitude. The K_m determination of testicular ALDH indicates that the CT fraction has the smallest apparent K_m, 0.118 mM, followed by the NC, 0.88 mM, and the MC fraction, 1.35 mM. The mito-

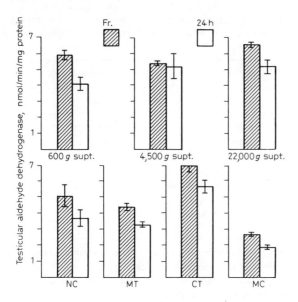

Fig. 4. The effect of storage at −20 °C on the specific activity of ALDH in various subcellular fractions and supernatant fluids of the rat testis. Values are means ± SE of specific activity of ALDH measured in the nuclear (NC), mitochondrial (MT), cytosolic (CT) and microsomal (MC) fractions of the rat testis. Enzymatic activity was also measured in the supernatant fluids of the various fractionation procedure. Each bar graph represents the mean ± SE of 5 independent determinations.

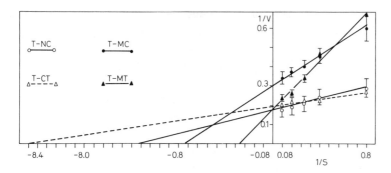

Fig. 5. Reciprocal plots of the velocity of ALDH reaction as a function of substrate (acetaldehyde) concentrations. Lineweaver-Burk plots were used for determinations of maximal velocity of the reaction (V_{max}) and the Michaelis-Menten constant (apparent K_m) of ALDH in the nuclear (NC), mitochondrial (MT), cytosolic (CT) and microsomal (MC) fractions of the rat testis. Each point is mean ± SE of 5–7 independent determinations.

Table II. The in vitro effects of some catecholamine metabolites, a histamine agonist and an antagonist and CNS stimulants on cytosolic alcohol (T-ADH) and aldehyde dehydrogenase (T-ALDH) of the rat testis preparation

Classification	Compound	Percent of controls	
		T-ADH	T-ALDH
Catecholamine metabolites	3-methoxytyramine	66.9 ± 6.8**	95.4 ± 4.1
	methanephrine	100.3 ± 1.4	102.9 ± 2.3
	vanillylmandelic acid	95.7 ± 3.7	98.8 ± 0.9
	homovanillic acid	91.7 ± 7.8	95.5 ± 5.6
Histamine agonist	histamine diphosphate	61.5 ± 2.5**	90.1 ± 6.0
antagonist	cimetidine	84.4 ± 4.0	96.1 ± 4.5
CNS stimulants	*d*-amphetamine	58.3 ± 5.7***	98.0 ± 9.3
	aminophylline	86.6 ± 11.2	95.0 ± 7.0

Compounds were dissolved in distilled water and added to the reaction mixture, 1.0 mM concentration. Values are means ± SE of percent changes of specific activity from mean control values (= 100%) obtained in the absence of the drug. Each value derives from a mean of 6–8 independent determinations. ** = $p < 0.02$; *** = $p < 0.01$.

chondrial ALDH showed the highest apparent K_m among testicular ALDH and hence the lowest affinity towards the substrate.

The in vitro effects of some selected catecholamine metabolites, histaminic compounds with agonist and antagonist properties and some CNS stimulants on cytosolic testicular ADH and ALDH are listed in table II. The results are expressed as percent changes in mean ± SE of specific activities of both enzymes as a function of the drugs used. They were added to the reaction mixture in equal molar concentration, 1.0 mM. The O-methylated metabolite of dopamine, 3-methoxytyramine, was the only compound of the catecholamine metabolites tested which inhibited testicular ADH. This inhibition amounted to approximately 33.1% ($p < 0.02$). There was 38.5% ($p < 0.02$) and a 15.6% ($p < 0.1$) in vitro reduction in specific activity of testicular ADH in the presence of 1 mM of histamine or 1 mM of cimetidine, respectively. A greater inhibition of testicular ADH occurred by *d*-amphetamine, approximately 42% ($p < 0.01$). Testicular ALDH showed little changes in specific activity as a function of the drugs used in vitro.

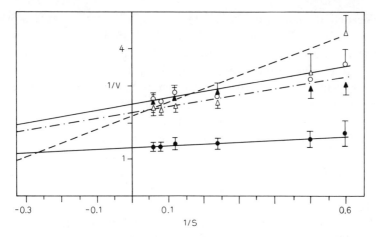

Fig. 6. Reciprocal plots of control and in vitro inhibition of rat cytosolic ADH.
3-Methoxytyramine (O), histamine diphosphate (△) or *d*-amphetamine (▲) were added to the reaction mixture in equimolar concentration (1.0 mmol/ml assay mixture). Each point is the mean ± SE of 5 independent determinations and the control (●) represents values obtained in the absence of drugs in reaction mixture.

Figure 6 shows the type of in vitro inhibition produced by $10^{-3}\,M$ of 3-methoxytyramine, of histamine diphosphate and of *d*-amphetamine and changes in testicular enzymatic ADH activity as related to substrate concentration using Lineweaver-Burk reciprocal plots in the absence of (control) and in the presence of 1 mM of these drugs. All these agents noncompetitively inhibited cytosolic testicular ADH.

Discussion

The present study confirms the initial finding [6] of the presence of ALDH in the CT fraction of the rat testis and extends it to report on the presence of NAD-linked ALDH in the NC, MT and in the MC preparations. This study also demonstrates that previously reported ADH, as assayed in total homogenates [10] or determined histochemically [2], is distributed between the NC and the CT fraction of the rat testis. The epididymal tissue contains ALDH only in the CT fraction. Moreover, no measurable ADH was detectable in the epididymis.

The activity measured in the various cellular components of the testis and in the CT fraction of the epididymis are of enzymatic source as evidenced by suppressed measurable activity at 4 °C and its deactivation by 50 °C. The results show that the enzymes measured are NAD-dependent and require sodium pyrophosphate buffer, pH 9.8, for maximal enzymatic activity. Furthermore, it appears that testicular ADH and epididymal ALDH are destroyed by storage overnight at −20 °C. This may be due to the presence of endogenous inhibitor, instability in the pH and/or of the buffer used. There was also a loss of enzymatic activity in testicular NC-ALDH by freezing at −20 °C with a minimal decrease in specific activity occurring in the CT fraction of the testis.

The kinetic portion of this study shows that the velocity of the ALDH reaction (V_{max}) is of the same order and magnitude in the NC, MT and CT preparations. However, a different pattern emerges from the determinations of the K_m value for testicular ALDH in the cellular components obtained. For example, the highest and lowest K_m values for testicular ALDH were measured in the MT and in the CT fraction, respectively. This suggests that the MT enzyme possesses the least affinity towards acetaldehyde as substrate. This is contrasted with greater affinity of cytosolic ALDH towards the same substrate. The order for increasing substrate affinity of testicular ALDH in various subcellular fractions is a follows: CT>NC>MC>MT.

The in vitro inhibition of cytosolic ADH by 3-methoxytyramine, the O-methylated metabolite of dopamine, and not by metanephrine, the O-methylated metabolite of epinephrine, may possibly indicate importance of the phenylethylamine moiety of the molecules for this action and/or may suggest the possible involvement of dopaminergic function in modulating the biotransformation of ET in the testis. The CNS stimulant *d*-amphetamine but not aminophylline noncompetitively inhibited testicular ADH. Both 3-methoxytyramine and *d*-amphetamine share a similar structure which may suggest the importance of the aromatic monoamine nucleus in the enzymatic inhibition observed. Cimetidine, a histaminergic H_2 receptor antagonist, and histamine, a compound with both H_1 and H_2 agonist histaminic properties, were effective in inhibiting testicular ADH in vitro.

The foregoing observation on the presence of ADH and ALDH in the testis suggests that the testis is an important endocrine gland in the detoxification of ET and of acetaldehyde. The wide subcellular distribution of both these enzymes suggests specific enzymatic compartmentalization which may be of future metabolic importance. Moreover, the difference in K_m values of the enzymes measured indicate the possible presence of various active forms of the enzymes, i.e. isoenzymes. This in turn may provide identification of substrate specificity and may conceivably lead to a better understanding of drug-drug interaction as related to activators, inhibitors and/or substrates specificity of these enzymes. Current experiments are in progress to evaluate some of these postulates.

Summary

The subcellular distribution of the enzymes primarily involved in the metabolism of ethanol and acetaldehyde were made in the rat testis and in the epididymis. The enzymes measured were alcohol dehydrogenase (ADH) and aldehyde dehydrogenase (ALDH). They were NAD-dependent, temperature sensitive and displaying maximal activity in the presence of pyrophosphate buffer at pH 9.8. Both enzymes were readily measurable in the 10% (w/v) testicular and epididymal homogenates. The ADH activity was mainly localized in the nuclear and in the cytosolic fractions of the testis compared to the absence of measurable ADH activity in the epididymis. Testicular ALDH was measurable in all subcellular preparations, i.e. in the nuclear, in the mitochondrial, in the cytosolic and in the microsomal fractions. Maximal testicular ALDH activity was determined in both the nuclear and the cytosolic components compared to a lower microsomal ALDH activity. Determination of K_m shows that cytosolic ALDH possesses the lowest apparent K_m as contrasted with a high value for the mitochondrial ALDH. Testicular cytosolic ADH but not ALDH was noncompetitively inhibited by 3-methoxytyramine, histamine and d-amphetamine in vitro.

References

1 Anderson, D.C.: The effect of alcohol on the hepatic metabolism of hormones. Eur. J. clin. Invest. 8: 267–268 (1978).
2 Ferguson, M.M.; Baillie, A.H.; Calman, K.C.; Heart, D.D.: Histochemical distribution of alcohol dehydrogenase in endocrine tissue. Nature, Lond. 210: 1277–1279 (1966).
3 Gerdes, H.: Alkohol und Endokrinum. Internist 19: 89–96 (1978).
4 Lieber, C.S.; De Carli, L.M.: Hepatic microsomal ethanol-oxidizing system: in vitro characteristics and adaptive properties in vivo. J. biol. Chem. 245: 2505–2512 (1970).
5 Lineweaver, H.; Burk, D.: Determination of enzyme dissociation constants. J. Am. chem. Soc. 56: 568–666 (1934).

6   Messiha, F.S.; Girgis, S.M.: Aldehyde dehydrogenase activity in the male rat reproductive tissue. Proc. west. Pharmacol. *21:* 353–356 (1978).
7   Messiha, F.S.; Tyner, G.S.; Girgis, S.M.: Property and specificity of aldehyde dehydrogenase in the rat testis; in Messiha, Tyner, Alcoholism: a perspective, pp. 467 to 477 (PJD Publ., New York 1980).
8   Messiha, F.S.: Testicular and epididymal aldehyde dehydrogenase in rodents: modulation by ethanol and disulfiram. Int. J. Androl. *3:* 375–382 (1980).
9   Messiha, F.S.; Hutson, J.: Distribution and kinetic properties of alcohol and aldehyde dehydrogenase in the rat testis. Archs Androl. (in press).
10   Van Thiel, D.H.; Gavaler, J.; Lester, R.: Ethanol inhibition of vitamin A metabolism in the testes: possible mechanism for sterility in alcoholics. Science *186:* 941–942 (1974).
11   Van Thiel, D.H.; Lester, R.: The effect of chronic alcohol abuse on sexual function. Clin. Endocrinol. Metab. *8:* 499–510 (1979).
12   Wright, J.: Endocrine effects of alcohol. Clin. Endocrinol. Metab. *7:* 351–373 (1978).

Dr. F.S. Messiha, Department of Pathology, Texas Tech University
Health Sciences Center, School of Medicine, Lubbock, TX 79430 (USA)

Epididymal Aldehyde Dehydrogenase: A Pharmacologic Profile[1]

F.S. Messiha[2]

Division of Toxicology, Department of Pathology, and Psychopharmacology Laboratory, Department of Psychiatry, Texas Tech University Health Sciences Center, School of Medicine, Lubbock, Tex., USA

Introduction

The presence of NAD-dependent aldehyde dehydrogenase (ALDH) in the rat epididymis has been the subject of several reports from this laboratory [2–6]. The present study evaluates some of the pharmacologic properties of this enzyme by studying the in vivo effect of various drugs on the specific activity of epididymal (EP)-ALDH and by determining some of the kinetics of this enzyme as related to certain drug action.

Material and Methods

Adult Sprague-Dawley rats were used throughout the experiments. They were 80 to 110 days old unless otherwise indicated, and were maintained on Purina pellet food and water ad libitum in a room with alternating 12 h dark and light cycles. Animals were sacrificed by decapitation and their epididymes dissected into the caput (CP), the cauda (CD) or used as a whole. Tissues were removed, weighed and homogenized in ice-cold 0.1 M KCl buffer, pH 6.8, to obtain a 10% (w/v) homogenate by a waring blender. The homogenate was centrifuged for 90 min at 22,000 g and the resulting cytoplasmic supernatant was used as the source for the EP-ALDH. In some experiments the testis was processed similar to the epididymis to obtain the cytosolic fraction. Aliquots of the cytosolic fraction were used for the protein determination by the biuret method and the enzyme activity was measured as described earlier [5].

---

[1] Supported in part by a grant from TTUHSC Biomedical Research Institute.
[2] The excellent technical assistance of Mr. *James Webb* is acknowledged.

Table I. Effect of various dosage and routes of administration of disulfiram (DIS) on endogenous rat EP-ALDH in the CP- and CD-epididymis of the rat

Experiment	Treatment group	DIS daily dose mg/kg	duration days	route	n	CP-ALDH nmol/min/mg protein	CD-ALDH
A	Controls	vehicle	2	i.p.	4	5.5 ± 0.5	7.2 ± 0.7
	DIS	6	2	i.p.	4	5.4 ± 0.6	7.0 ± 0.4
B	Controls	vehicle	7	i.p.	7	4.4 ± 0.3	5.6 ± 0.4
	DIS	25	7	i.p.	7	4.0 ± 0.6	4.3 ± 0.3
C	Controls	vehicle	4	p.o.	5	8.3 ± 0.9	7.4 ± 0.3
	DIS	400	4	p.o.	5	2.8 ± 0.5**	3.8 ± 0.8*
D	Controls	–	61	food	10	9.6 ± 0.7	9.8 ± 1.8
	DIS	15	61	food	10	7.0 ± 0.5*	4.7 ± 0.5**

** $p<0.001$, * $p<0.01$.

Values are means ± SE of specific activity (nmol/min/mg protein) of the number of animals given in parentheses. DIS was dissolved in an organic solvent and suspended into drake oil by evaporation over a water bath. Identical amounts of organic solvents were similarly treated with drake oil in the absence of the drug and used as the vehicle for the controls of experiments A, B and C. In experiment D, DIS was mixed with the food (see methods).

In the first set of experiments, the effect of pyrazole (PYZ) on the endogenous specific activity of ALDH in the CP and in the CD epididymis was studied as a function of the following experimental conditions. (a) PYZ time course: PYZ was administered, 3 mmol/kg i.p., and the animals were sacrificed at various time intervals between the initial hour and the 16th h of the PYZ injection; (b) PYZ dose response: PYZ was injected, 2, 5 or 8 mmol/kg i.p., and the animals were sacrificed 1 h after the PYZ treatment; (c) PYZ-ethanol interaction: PYZ was administered, 5 mmol/kg i.p., to rats maintained on water or on 10% ethanol solution as the only drinking fluid and were pair fed for the duration of the 14-day experiment, and (d) PYZ short-term effect: PYZ was injected, 100 mg/kg i.p., once daily for 7 consecutive days and the animals were sacrificed 16 h after the terminal injection. PYZ was dissolved in saline and the corresponding controls received the vehicle, physiological saline.

In the second set of experiments, the effect of various doses of disulfiram (DIS), routes and duration of the DIS administration on CP-ALDH and CD-ALDH were studied. DIS was injected, 6 mg/kg i.p., once daily for 2 consecutive days, 25 mg/kg i.p. once daily for 7 days, given orally in a massive daily dose, 400 mg/kg for 4 consecutive days, or administered in food by mixing DIS with the Purina food on the basis of a daily dose delivery of 15 mg/kg/day for 61 days. In the latter experiment, DIS was dissolved in an organic solvent, mixed with the food and the organic solvent evaported in a stream of cold air. The controls received identical amounts of food (25 g/day), which had been treated similarly with the organic solvent but in the absence of DIS. In all other experiments, DIS was dissolved in an organic solvent, suspended in vegetable oil and the organic solvent evaporated over water bath. Equal volumes of the vehicle, which had been treated similarly but without the DIS addition, were given to the controls (0.05 ml/100 g body weight).

In the third set of experiments, the effects of short-term administration of selected pharmacologic agents on endogenous epididymal and testicular (T) cytosolic ALDH was studied. Depending on the solubility of the agents used, they were dissolved either in saline or in organic solvents and suspended into vegetable oil prior to evaporating the organic solvents over a water bath. Controls received the vehicle used for drug preparation. All drugs were administered intraperitoneally, and the animals were sacrificed by decapitation 1 h subsequent to the final injection schedule indicated in table II. The drugs used were of analytical grade and are all listed in table II.

In the fourth set of experiments, the effect of short-term administration of certain steroids and a nonsteroidal antiandrogenic compounds on cytosolic EP-ALDH and T-ALDH was studied in the rat. These agents are listed in table III. They were dissolved in an organic solvent and suspended in vegetable oil and treated as described above, prior to their daily oral administration of the doses and for the duration of time indicated in table III.

The results are expressed as means ± SE of specific activity (nmol/min/mg protein). Student's t test for independent means was used for the statistical evaluation of the data.

Results

Figure 1 shows the relationship between the PYZ dose administered (lower panel) and the duration of action of PYZ (upper panel) on endogenous EP-ALDH. The time course study for PYZ indicates that administration of a single dose of PYZ, 3 mmol/kg i.p., to adult male rats exerted little effect on EP-ALDH during the initial period of the PYZ injection. However, a transient increase in CP-ALDH by PYZ was evident during the 8th h of drug administration from saline controls ($p<0.01$). This is contrasted with a moderate but not significant inhibition of CP-ALDH occurring 16 h after the PYZ injection ($p<0.1$). The effect of increasing doses of PYZ on EP-ALDH shows that administration of a single dose of PYZ as high as 5 mmol/kg i.p. resulted in inhibition of ALDH in the CP portion of the epididymis 1 h after drug injection. A massive dose of PYZ, 8 mmol/kg i.p., produced a marked inhibition on CP-ALDH ($p<0.001$) and CD-ALDH ($p<0.01$) at identical time intervals.

Figure 2 shows the effect of a single dose of PYZ, 5 mmol/kg i.p., on CP- and CD-ALDH in rats maintained on water or 10% ethanol

Table II. The effect of short-term administration of some psychoactive drugs, metabolic inhibitors and other pharmacologic agents on endogenous cytosolic EP- and T-ALDH in the rat

Experiment	Drug classification	Treatment compound	dose mg/kg/day	duration days	n	EP-ALDH nmol/min/mg protein	T-ALDH
I	Controls	vehicle	–	6	6	6.4 ± 0.3	3.5 ± 0.5
	Antianxiety	diazepam	5	6	6	6.0 ± 0.8	3.6 ± 0.3
	Antidepressant	imipramine	10	6	6	6.2 ± 1.0	4.1 ± 0.9
	Antipsychotics	chlorpromazine	10	6	6	7.1 ± 0.9	3.4 ± 0.3
		haloperidol	3	8	8	8.2 ± 2.1	3.1 ± 0.4
II	Controls	vehicle	–	7	6	6.9 ± 0.5	3.7 ± 0.3
	Antiparkinsonian drugs	apomorphine	3	7	6	7.3 ± 0.6	3.7 ± 0.5
		amantadine · HCl	100	7	5	7.0 ± 0.6	3.3 ± 0.4
		benzotropine mesylate	0.02	7	5	6.6 ± 0.7	4.0 ± 0.3
		cyclobenzaprine	50	7	5	7.8 ± 1.3	4.1 ± 0.4
		trihexyphenidyl · HCl	0.10	7	5	7.6 ± 0.8	4.4 ± 0.4
III	Controls	vehicle	–	10	6	10.5 ± 0.6	3.2 ± 0.3
	Antiparkinsonian drugs	L-dopa methylester	200	10	5	10.6 ± 1.0	3.5 ± 0.1
		carbidopa	100	10	4	6.9 ± 1.2***	2.4 ± 0.2*
		L-dopa + carbidopa	200 + 100	10	4	6.5 ± 1.9**	2.2 ± 0.4
IV	Controls	vehicle	–	6	6	7.8 ± 1.3	5.0 ± 0.4
	Metabolic inhibitors	iproniazide (MAOI)	50	6	5	11.3 ± 0.7*	4.9 ± 0.4
		p-chlorphenylalanin methylester	130	6		8.9 ± 1.8	6.0 ± 0.5

Table II. (continued)

Experiment	Drug classification	Treatment compound	dose mg/kg/day	duration days	n	EP-ALDH nmol/min/mg protein	T-ALDH
V	Controls	vehicle	–	6	6	9.8 ± 1.2	4.7 ± 0.6
	CNS stimulants	aminophylline	50	6	6	10.0 ± 1.0	4.7 ± 0.4
		caffeine · HCl	50	6	6	11.0 ± 1.2	3.8 ± 0.4
VI	Controls	vehicle	–	6	5	11.6 ± 1.0	6.7 ± 0.6
	Cholinomimetics	choline/HCl	75	6	4	8.7 ± 0.9	6.5 ± 0.2
		physostigmine	0.5	1	6	11.4 ± 1.6	4.9 ± 0.3*
		pilocarpine	25	6	4	8.4 ± 1.8	7.4 ± 2.0
VII	Controls	vehicle	–	6	6	16.1 ± 1.3	5.0 ± 0.2
	Vitamin A and its aldehyde	retinol	500	6	6	14.2 ± 0.7	4.7 ± 0.4
		retinal	5	6	6	13.4 ± 0.7*	4.9 ± 0.2

*** $p < 0.02$; ** $p < 0.05$; * not significant ($p < 0.1$).

Drugs were dissolved in saline or in ethylacetate. The latter was suspended in vegetable oil and the organic solvent evaporated over a water bath. The controls received the respective vehicle, saline or vegetable oil. Drugs were injected intraperitoneally in the dose given once daily and for the duration of time indicated. Animals were sacrificed 1 h subsequent to the terminal dosage. Values are for means ± SE of specific activity for the number of animals given in parentheses. A total of 154 rats were used.

Table III. The effect of short-term administration of steroidal hormones and of a nonsteroidal antiandrogen on cytosolic NAD-dependent EP- and T-ALDH in the rat

Experiment	Drug classification	Treatment compound	dose mg/day	duration days	n	EP-ALDH nmol/min/mg protein	T-ALDH
I	Control	vehicle*	–	8	6	11.6 ± 1.0	5.9 ± 0.6
	Androgen	testosterone	2	8	6	12.3 ± 1.5	5.3 ± 0.5
II	Control	vehicle	–	6	13	11.3 ± 2.2	4.5 ± 0.3
	Antiandrogen	flutamide	10	6	12	9.7 ± 1.0	5.2 ± 0.4
		SCH-16423	10	6	12	12.7 ± 1.0	5.9 ± 0.5*
III	Control	vehicle	–	5	7	11.8 ± 1.1	6.7 ± 0.1
	Progesterone	progesterone	10	5	6	11.0 ± 1.5	6.9 ± 0.3
	Progesterone (synth.)	norethindrone acetate	0.1	5	5	12.8 ± 1.4	7.6 ± 0.7
	Estrogen	estradiol	2	5	7	12.3 ± 1.7	6.1 ± 0.7
	Estrogen (synth.)	ethynyl estradiol	0.1	5	7	11.9 ± 1.4	5.0 ± 0.3*

* $p < 0.01$.

[1] Drugs were dissolved in the vehicle (see 'Methods'), vegetable oil, and were administered orally in a volume not greater than 0.2 ml. Values are means + SE of specific activity for number of animals given in parentheses.

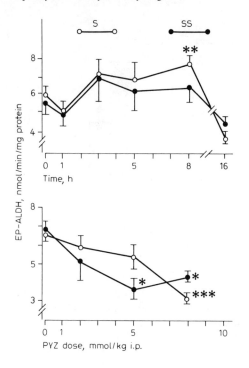

Fig. 1. The in vivo effect of PYZ on endogenous specific activity of rat ALDH in the caput (S) and in the cauda (SS) epididymis as a function of time and the PYZ dose given.

PYZ was injected, 3 mmol/kg i.p., and the rats were sacrificed at the time intervals indicated (upper panel). PYZ was also administered, in various dosages, 2, 5 or 8 mmol/kg i.p., and the animals were sacrificed 1 h after the time of the PYZ treatment (lower panel). The controls received the vehicle, physiological saline, i.p. Each point represents means ± SE of specific activity of ALDH (nmol/min/mg protein) of 4–5 independent determinations. A total of 48 rats were used. *** $p<0.001$, ** $p<0.01$, * $p<0.02$.

solution as the sole drinking fluid for 14 consecutive days. Ethanol-drinking rats were pair fed with the water-drinking animals. All animals were sacrificed 16 h after the PYZ injection. The specific activity of ALDH was decreased subsequent to the PYZ injection in the CP portion of the epididymis from corresponding controls in animals maintained on water ($p<0.01$) or on ethanol ($p<0.05$). PYZ inhibitory action on CD-ALDH was greater than that determined for the CP-ALDH of rats maintained on water ($p<0.001$) or on the 10% ethanol solution ($p<0.05$).

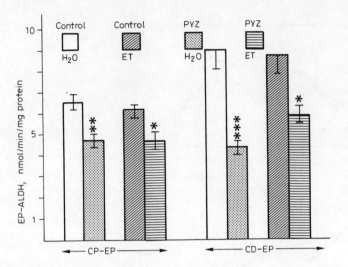

Fig. 2. The effect of single dose of PYZ on endogenous specific activity of ALDH in the CP-EP and the CD-EP as a function of drinking fluid. Animals were maintained on water or 10% (w/w) ethanol (ET) solution as the sole drinking fluid and were pair fed. They were sacrificed 16 h after administration of a single dose of PYZ, 5 mmol/kg i.p. Each bar represents mean ± SE derived from 12 rats for each treatment modality. *** $p<0.001$, ** $p<0.01$, * $p<0.05$.

In a separate experiment the effect of short-term administration of PYZ on CP- and CD-ALDH was studied. Injection of PYZ, 100 mg/kg i.p., once daily for 7 consecutive days inhibited CP-ALDH to 4.7 ± 0.3 nmol/min/mg protein compared to mean saline controls of 6.5 ± 0.4 units ($p<0.01$). Concomitantly, a greater inhibition of CD-ALDH by PYZ was noted. This is indicated by 5.4 ± 0.5 nmol/min/mg protein, determined for the specific activity of CD-ALDH derived from the PYZ-treated rats compared to 9.1 ± 0.9 units for the respective controls ($p<0.001$).

The effect of various doses of DIS, their routes and duration of administration of DIS on endogenous EP-ALDH in the CP and CD epididymis is summarized in table I. Intraperitoneal injections of a small daily dosage of DIS, 6 or 25 mg/kg, for the respective 2 or 7 consecutive days did not alter the specific activity of EP-ALDH compared to controls. Oral administration of a massive daily large dose of DIS, 400 mg/kg, for 2 days exerted marked and significant effects on both parts of the epididymis measured for ALDH (from $p<0.01$ to $p<0.001$). Administration of a small dose of DIS but for a longer period of time, i.e., 15 mg/kg/day for 61 days, was required for significant inhibition of CP- and CD-ALDH ($p<0.01$ and $p<0.001$).

Table II lists the effect of various pharmacologic agents tested on cytoplasmic EP- and T-ALDH in the rat. Intraperitoneal injections of the daily dose of some selected psychoactive agents shown for the duration of time indicated did not alter the specific activity of cytosolic ALDH in the whole epididymis or in the testis significantly from respective controls. Similarly, the antiparkinsonian medication tested was devoid of action on ALDH in the testicles. Some of the metabolic inhibitors selected produced some changes in endogenous EP-ALDH. For example, iproniazide, a monoamine oxidase inhibitor, enhanced the specific activity of endogenous ALDH in the epididymis by approximately 14.5% compared to controls without a concomitant change in T-ALDH. However, this induction of EP-ALDH was not statistically significant ($p<0.1$). Short-term administration of carbidopa, 100 mg/kg/day for 10 consecutive days, an inhibitor of aromatic L-amino acid decarboxylase, inhibited EP-ALDH by approximately 34% compared to corresponding controls ($p<0.02$). Physostigmine was the only cholinomimetic drug tested which altered T-ALDH but not EP-ALDH compared to controls after single dose administration ($p<0.1$).

The effect of short-term oral administration of certain androgens, antiandrogen, estrogenic and progesteron-like compounds on cytoplasmic ALDH in the whole rat epididymis and testis is given in table III. Administration of testosterone, 2 mg/kg/day for 8 consecutive days, did not alter the specific activity of ALDH in the tissues studied. This is compared with 13% ($p<0.01$) induction of T-ALDH compared to controls by the active metabolite of the nonsteroidal antiandrogen, i.e., SCH-16423. Flutamide administration increased T-ALDH by approximately 11.6% compared to controls, which was not statistically significant ($p<0.1$). Administration of progesterone or norethindrone acetate, a synthetic progesterone-like compound, exerted little action on ALDH from the respective controls as does the administration of a large dose of estradiol. Oral administration of norethynyl estradiol, 100 μg/kg/day for 5 consecutive days, resulted in inhibition of T-ALDH by 25% compared to respective controls ($p<0.01$).

Discussion

This study shows that both PYZ and DIS, the respective inhibitors of hepatic alcohol dehydrogenase, and ALDH, are inhibitors for EP-ALDH in vivo. It has been shown in a pilot study that PYZ did not alter the specific activity of EP-ALDH in vitro. Therefore, the observed PYZ effect may have been due to biologically active metabolite(s) of PYZ. DIS inhibited EP-ALDH when given in a massive dose for a short period of time or when administered in a smaller daily dosage semichronically. The inhibition of EP-ALDH by DIS will result in accumulation of and in a build-up of ethanol-derived acetaldehyde in the testicles which may conceivably provide a new experimental approach for evaluating the toxicity of DIS on the gonads.

ALDH is widely distributed in extracerebral tissues of mammals, and rat liver ALDH, which is the most studied organ, is altered in vivo by drugs which possess various pharmacologic properties, suggesting a wide spectrum of specificity of hepatic ALDH towards substrates and inhibitors. However, the present results suggest that EP-ALDH may possess greater specificity than hepatic ALDH. This is indicated by the lack of response of EP-ALDH to short-term administration of a large number of drugs with different pharmacologic profiles, some of which are potent modifiers of hepatic ALDH in vivo [7–12]. It is conceivable that the lack of effect of the agents studied on endogenous EP-ALDH may be ascribed to their extent of penetration into the epididymis and/or their metabolism to physiologically innert metabolites. However, it appears that T-ALDH is more sensitive to pharmacologic interventions than EP-ALDH.

The efficacy of the biologically active metabolite of flutamide, a nonsteroidal antiandrogen, and the estrogenic compound used in altering specific activity of endogenous EP-ALDH is of particular interest. Administration of flutamide, where clinically indicated, may result in enhanced metabolism of exogenously consumed ethanol-derived acetaldehyde by induction of T-ALDH and thus minimizes from ethanol-flutamide interaction in the testis. Both flutamide and SCH-16423 have been shown to be devoid of action on hepatic alcohol dehydrogenase and ALDH when administered in larger daily dosages than that used in this study [13]. The inhibitory action of the estrogenic compound used on T-ALDH and its inhibition of rat liver ALDH in vivo [14] may have some clinical implications. For example, excessive consumption of alcoholic beverages during treatment of metastatic carcinoma of the prostate by estrogens and the accidental contact with estrogen containing medications by man [1] should be contraindicated.

The foregoing observations on EP-ALDH calls for more detailed studies on the functional activity and specificity of this enzyme.

Summary

The in vivo effect of various agents, with different pharmacologic properties, on cytosolic EP-ALDH was studied in the rat. Injection of a large dose of PYZ inhibited EP-ALDH 16 h after the injection. Short-term administration of a moderate daily dosage of PYZ also inhibited endogenous EP-ALDH from saline controls. Semichronic administration of a small daily dosage of DIS resulted in inhibition of EP-ALDH. Short-term administration of drugs with various pharmacologic profile produced little change in the specific activity of EP-ALDH. This is compared with an induction and an inhibition of T-ALDH as a function of treatment with an antiandrogen and an estrogenic drug, respectively. The results show lack of response of EP-ALDH towards pharmacologic intervention which may imply certain specificity for EP-ALDH.

References

1 Diraimondo, C.V.; Roach, A.C.; Meador, C.K.: Gynecomastia from exposure to vaginal estrogen cream. New Engl. J. Med. *302:* 1089–1099 (1980).
2 Messiha, F.S.; Girgis, S.M.: Aldehyde dehydrogenase activity in the male rat reproductive tissues. Proc. west. Pharmacol. *21:* 353–356 (1978).
3 Messiha, F.S.: Epididymal alcohol and aldehyde dehydrogenase in rodents. Proc. west. Pharmacol. *22:* 133–136 (1979).
4 Messiha, F.S.; Tyner, G.S.; Girgis, S.M.: Aldehyde dehydrogenase in the rats reproductive organs; in Messiha, Tyner, Alcoholism: a perspective, pp. 467–477 (PJD Publ., New York 1980).
5 Messiha, F.S.: Testicular and epididymal aldehyde dehydrogenase in rodents: modulation by ethanol and disulfiram. Int. J. Androl. *3:* 375–382 (1980).
6 Messiha, F.S.; Hutson, J.: Distribution and kinetic properties of alcohol and aldehyde dehydrogenase in the rat testis. Archs Androl. *6:* 243–248 (1981).
7 Messiha, F.S.: Antagonism of ethanol-evoked responses by amantadine: a possible clinical application. Pharmacol. Biochem. Behav. *8:* 573–577 (1978).
8 Messiha, F.S.: Modulation of hepatic aldehyde dehydrogenase by carbidopa. Res. Commun. chem. Pathol. Pharmacol. *20:* 601–604 (1978).
9 Messiha, F.S.: Voluntary drinking of ethanol by the rat: biogenic amines and possible underlying mechanism. Pharmacol. Biochem. Behav. *9:* 379–384 (1978).
10 Messiha, F.S.; Barnes, C.D.: Cyclobenzaprine and ethanol interaction. Pharmacol. Biochem. Behav. *10:* 947–949 (1979).

11 Messiha, F.S.; Hughes, M.J.: Liver alcohol and aldehyde dehydrogenase: inhibition and potentiation by histamine agonists and antagonists. Clin. exp. Pharmacol. Physiol. *6:* 281–292 (1979).
12 Messiha, F.S.; Sabonghy, M.M.: Cholinergic, anticholinergic agents and ethanol metabolizing enzymes. Arch. int. Pharmacodyn. Ther. *248:* 4–12 (1980).
13 Messiha, F.S.: Androgens, antiandrogens and voluntary intake of ethanol by the male rat. Res. Commun. Subst. Abuse *1:* 1–8 (1980).
14 Messiha, F.S.; Lox, C.D.; Heine, W.: Studies on ethanol and oral contraceptives: feasibility of a hepatic-gonadal link. Res. Commun. Subst. Abuse *1:* 315–333 (1980).

Dr. F.S. Messiha, Department of Pathology, Texas Tech University,
Health Sciences Center, School of Medicine, Lubbock, TX 79430 (USA)

Behavioral Pharmacology

Prog. biochem. Pharmacol., vol. 18, pp. 179–189 (Karger, Basel 1981)

Voluntary Ethanol Consumption by Female Offspring from Alcoholic and Control Sinclair(S-1) Miniature Dams[1]

Myron E. Tumbleson, James D. Dexter, Charles C. Middleton

Sinclair Comparative Medicine Research Farm, College of Veterinary Medicine, and School of Medicine, University of Missouri, Columbia, Mo., USA

Introduction

Sullivan [12] reported that maternal inebriety had deleterious effects on offspring and that removal of female drunkards prevents procreation of children most apt to be a burden to society. Recently, *El-Guebaly and Offord* [3] and *Goodwin* [5] reviewed the literature with respect to alcoholic offspring. As reported earlier that maternal inebriety resulted in child neglect [12], *Chafetz* et al. [2] suggested that children of alcoholic parents had increased chances of incurring serious illness or accident which was related to parental absence or neglect.

The correlation of alcoholism in adoptees with an alcoholic history in the biologic parents was suggestive of the importance of a genetic factor [1]. Also, it has been reported [6, 11] that adoptees from biologic alcoholic parents had a greater history of drinking problems than did controls. *Fine* et al. [4] indicated that regardless of specific etiologic factors, the presence of parental alcoholism introduces a serious deterrent for healthy child development. *MacKay* [9] found a higher percentage of addictive drinkers with frequent parental alcoholism. However, *Roe* [10] and *Goodwin* et al. [7] reported that alcoholic offspring raised by foster parents were similar to controls and did not create a risk of depression.

The study reported herein was designed to evaluate: (a) meal drinking by dams, (b) effect of maternal alcoholism on size and number

---

[1] Supported in part by a grant from the United States Brewers Association.

of offspring, (c) voluntary ethanol consumption during gestation and lactation, (d) voluntary ethanol consumption by offspring from alcoholic and control miniature swine dams, and (e) individual variability of voluntary ethanol consumption.

Material and Methods

8–14 weeks prior to breeding, 12 yearling Sinclair(S-1) miniature gilts were selected from the breeding herd and assigned randomly to treatment. 6 gilts were pen housed and fed a control diet, from individual stalls, once daily. The 6 experimental dams were pen housed and fed, from individual stalls, once daily. Each of the experimental dams was allowed access to dietary ethanol for two 1-hour periods daily, from 08.00 to 09.00 and 16.00 to 17.00. All animals were provided fresh drinking water ad libitum.

Dietary ethanol was supplied from lixits in the individual stalls. Ethanol was supplied as a 20% (w/v) aqueous solution. Individual body weights were recorded biweekly. Rations were adjusted weekly to effect similar intakes per pig per day of calories, protein, vitamins and minerals. Voluntary ethanol consumption was quantitated daily, between 08.00 and 09.00 h, from 4 weeks prebreeding until 5 weeks postweaning.

Within the first 16 h postfarrowing, each piglet was weighed, the needle teeth were clipped and body length (snout-rump) was determined. Body weights were recorded weekly until 40 weeks of age. Litters were kept intact until weaning at 5 weeks of age. At 1 week of age, piglets were provided with creep feed. Ethanol lixits were placed in a position where only the dams had access to ethanol. Piglets had fresh water ad libitum. Therefore, the only source of dietary ethanol for the piglets was from mothers milk. At 5 weeks of age, females were pen housed with 5–7 pigs per pen as were males. Until 17 weeks of age, all piglets received feed and water ad libitum.

At 12 weeks postweaning, 7 gilts from alcoholic mothers and 9 gilts from control mothers were selected to assess voluntary ethanol consumption of alcoholic versus control offspring. Each pig was housed individually with chain link fence between 2×2 m pens with concrete floors. Fresh drinking water was supplied ad libitum. Each pig was given a daily ration of feed to provide similar quantities of calories, protein, vitamins and minerals per pig per day. Daily ethanol consumptions were recorded between 08.00 and 09.00 h. During weeks 1 and 2, 3 and 4 and 5 through 23 on test, dietary ethanol was provided as a 10, 15 and 20% (w/v) aqueous solution.

Results

Voluntary ethanol consumption (fig. 1) averaged 1.83 g/kg body weight/day during the first 14 weeks of the 16-week gestation period. During the last 2 weeks of gestation and the first 2 weeks postpartum, voluntary ethanol consumption was decreased; whereas, there was an increased consumption postweaning. 3 of the 6 dams consumed similar

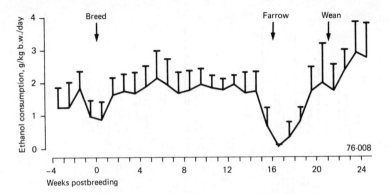

Fig. 1. Mean (±SD) voluntary ethanol consumptions (g/kg body weight/day) computed weekly for 6 dams from 4 weeks prebreeding through 5 weeks of lactation. Each dam was allowed access to ethanol for two 1-hour periods daily.

Table I. Ethanol consumption from 4 weeks prebreeding to 1 week prepartum

Dam No.	%		g/kg body weight		
	a.m.	p.m.	a.m.	p.m.	day
1163	33	67	0.42	0.87	1.3
1188	22	78	0.29	1.01	1.3
1172	53	47	0.91	0.80	1.7
1173	50	50	0.89	0.91	1.8
1157	52	48	1.00	0.91	1.9
1156	41	59	0.91	1.32	2.2

amounts of ethanol at the morning and afternoon feedings (table I); however, the other 3 dams drank 59–78% at the afternoon feeding.

Ethanol consumption per feeding is depicted in figures 2–7 for individual pigs. Dams 1163, 1188, 1172, 1173, 1157 and 1156 consumed in excess of 1.8 g ethanol/kg body weight/feeding 17, 21, 15, 13, 1 and 24 times, respectively, and less than 0.1 g/kg body weight/feeding 73, 67, 25, 18, 6 and 15 times, respectively, during the first 216 feedings of gestation.

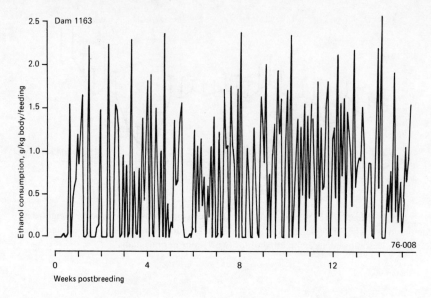

Fig. 2. Voluntary ethanol consumption, during the first 15 weeks of gestation, by dam 1163. Consumption was recorded for each of two 1-hour feedings daily.

There were no differences in gestation length, number of piglets born or body length of newborn piglets from alcoholic or control dams (table II). Newborn piglets from control dams weighed 20% more than piglets from alcoholic dams, i.e., 711 vs. 590 g, respectively.

Beginning at 17 weeks of age (12 weeks postweaning), each of the 16 offspring was given dietary ethanol ad libitum for 23 weeks. Mean body weights at birth, weaning, initiation of voluntary ethanol consumption and after 23 weeks on test are listed in table III and depicted graphically in figure 8.

During the first 15 weeks on test, alcoholic offspring consumed 7.7 g ethanol/kg body weight/day; whereas, offspring from control dams consumed only 5.6 g ethanol/kg body weight/day (table IV, fig. 9). For the last 8 weeks on test, voluntary ethanol consumptions were similar, at 7.2 g ethanol/kg body weight/day, for gilts from the 2 groups. Percentages of calories from ethanol, from 16 through 23 weeks on test, were 49 and 51 for control and alcoholic offspring, respectively (fig. 10).

Alcoholic Offspring 183

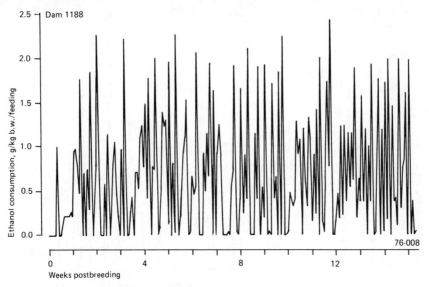

Fig. 3. Voluntary ethanol consumption, during the first 15 weeks of gestation, by dam 1188. Consumption was recorded for each of two 1-hour feedings daily.

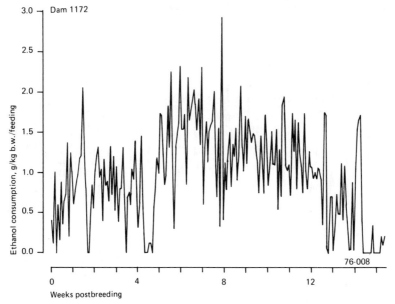

Fig. 4. Voluntary ethanol consumption, during the first 15 weeks of gestation, by dam 1172. Consumption was recorded for each of two 1-hour feedings daily.

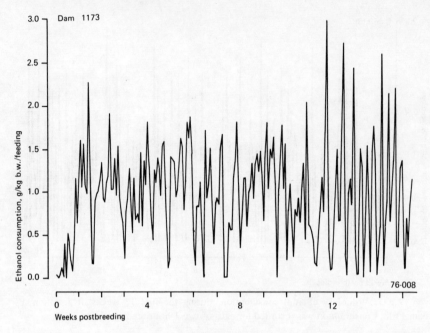

Fig. 5. Voluntary ethanol consumption, during the first 15 weeks of gestation, by dam 1173. Consumption was recorded for each of two 1-hour feedings daily.

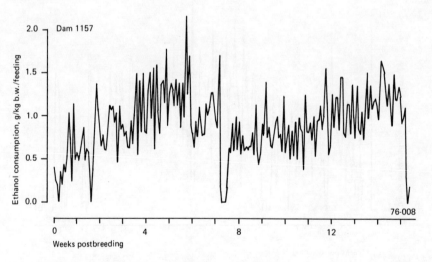

Fig. 6. Voluntary ethanol consumption, during the first 15 weeks of gestation, by dam 1157. Consumption was recorded for each of two 1-hour feedings daily.

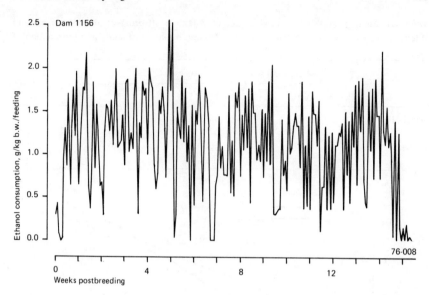

Fig. 7. Voluntary ethanol consumption, during the first 15 weeks of gestation, by dam 1156. Consumption was recorded for each of two 1-hour feedings daily.

Table II. Gestation periods, numbers of piglets born and piglet sizes

Dam No.	Alcoholic vs. control	Gestation period days	Piglets born, n	Birth weight, g	Body length, mm
1163	A	115	7	719 ± 101	160 ± 5
1188	A	113	6	527 ± 57	154 ± 11
1172	A	117	6	495 ± 94	154 ± 9
1173	A	112	6	685 ± 82	170 ± 14
1157	A	114	6	523 ± 102	148 ± 20
1156	A	118	3	553 ± 29	150 ± 9
Mean	A	115	5.7	590	156
1147	C	115	8	710 ± 119	162 ± 16
1177	C	118	7	810 ± 120	169 ± 12
1169	C	111	7	643 ± 86	154 ± 8
1174	C	116	4	845 ± 21	169 ± 2
1148	C	115	4	563 ± 147	154 ± 19
1124	C	116	2	630 ± 71	162 ± 11
Mean	C	115	5.3	711	162

Table III. Mean (±SD) body weights (kg) of 7 female offspring of alcoholic dams and 9 female offspring of control dams at birth, weaning, initiation of study and termination of study

	Age, weeks			
	birth	5	17	40
Alcoholic	0.637	3.00	7.43	20.4
	0.117	0.73	1.99	1.7
Control	0.767	3.88	7.89	22.0
	0.139	0.45	2.84	2.3

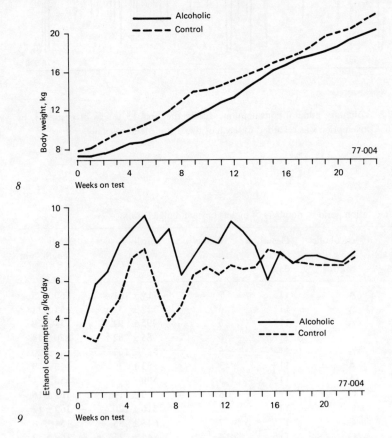

Fig. 8. Mean body weights of 7 female offspring of alcoholic dams and 9 female offspring of control dams. Each pig was 17 weeks of age at the initiation of the study.

Fig. 9. Mean weekly ethanol consumption (g/kg/day) by 7 female offspring of alcoholic dams and 9 female offspring of control dams.

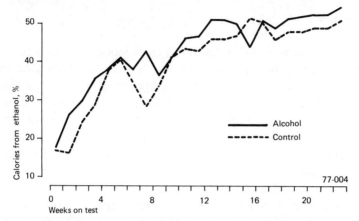

Fig. 10. Mean percent calories from ethanol for 7 female offspring of alcoholic dams and 9 female offspring of control dams.

Table IV. Mean voluntary ethanol consumption (g/kg/day) by 7 female offspring of alcoholic dams and 9 female offspring of control dams; intakes were recorded from 17 to 40 weeks of age

Gilt No.	Alcoholic vs. control	Weeks on test		
		1–15	16–23	1–23
4174	C	4.02	3.12	3.7 ± 1.6[1]
4178	C	4.11	5.24	4.5 ± 1.1
4173	C	3.64	6.46	4.6 ± 1.9
4092	A	5.11	3.83	4.7 ± 1.3
4186	C	4.96	6.90	5.6 ± 1.9
4105	C	5.49	7.21	6.1 ± 2.4
4106	C	5.22	8.53	6.4 ± 2.6
4112	A	7.22	5.68	6.7 ± 1.6
4117	C	7.02	6.55	6.9 ± 2.3
4097	A	6.39	7.79	6.9 ± 2.5
4111	A	7.00	7.11	7.0 ± 1.6
4175	C	6.79	9.29	7.7 ± 3.4
4100	A	7.81	8.34	8.0 ± 2.9
4101	A	11.02	7.19	9.7 ± 3.4
4098	A	9.48	10.34	9.8 ± 2.0
4177	C	9.36	11.18	10.0 ± 2.8
	alcoholic	7.72	7.18	7.5
	control	5.62	7.16	6.2
All pigs		6.54	7.17	6.8

[1] Standard deviations of weekly mean consumptions.

Discussion

Little et al. [8] reported a decreased consumption of alcoholic beverages in women during pregnancy with the subjects indicating that alcohol had become disagreeable and unappealing. During the latter stages of gestation, ethanol consumption by miniature swine dams also was lower. Whether this was due to physiologic, psychologic or teleologic reasons was not evaluated in our study. During and subsequent to lactation, voluntary ethanol consumption increased; therefore, the change perhaps was related to parturition.

The 2 dams (1163 and 1188) with the lowest mean daily consumptions of ethanol had numerous feedings when small quantities were ingested. However, other than dam 1157, the numbers of times when more than 1.8 g ethanol/kg body weight/feeding were similar. When depicting ethanol consumption/feeding, the dissimilarities in intakes during morning and afternoon feedings must be considered. However, consumption during a given 1-hour period may be more important physiologically than mean daily intake.

Mean body weights of selected female offspring from alcoholic and control miniature swine dams were not different mathematically; however, the slightly lower mean weights of alcoholic offspring continued throughout the duration of the study. Mean ethanol consumption of alcoholic offspring was greater than for control offspring for the first 15 weeks on test. As the females and males from the experimental and control dams had been housed together on a random basis, depending upon weaning date, they had been given the same diets for a period of 12 weeks prior to evaluation of voluntary ethanol consumption determinations. Therefore, we conclude that differences (2 g ethanol/kg body weight/day) in voluntary ethanol consumptions were due to fetal and neonatal environments.

Summary

At 17 weeks of age, 12 weeks postweaning, 7 gilts from alcoholic dams and 9 gilts from control dams were given ethanol ad libitum. For the first 15 weeks on test, alcoholic and control offspring consumed 7.7 and 5.6 g ethanol/kg body weight/day, respectively. From 16 through 23 weeks on test, alcoholic and control offspring consumed 7.2 and 7.2 g ethanol/kg body weight/day, respectively, which averaged 50% of caloric intake.

References

1 Cadoret, R.J.; Gath, A.: Inheritance of alcoholism in adoptees. Br. J. Psychiat. *132:* 252–258 (1978).
2 Chafetz, M.E.; Blane, H.T.; Hill, M.J.: Children of alcoholics. Observations in a child guidance clinic. Q. Jl Stud. Alcohol *32:* 687–698 (1971).
3 El-Guebaly, N.; Offord, D.R.: The offspring of alcoholics: a critical review. Am. J. Psychiat. *134:* 357–365 (1977).
4 Fine, E.W.; Yudin, L.W.; Holmes, J.; Neinemann, S.: Behavorial disorders in children with parental alcoholism. Ann. N.Y. Acad. Sci. *273:* 507–517 (1976).
5 Goodwin, D.W.: Alcoholism and heredity. A review and hypothesis. Archs gen. Psychiat. *36:* 57–61 (1979).
6 Goodwin, D.W.; Schulsinger, F.; Hermansen, L.; Guze, S.B.; Winokur, G.: Alcohol problems in adoptees raised apart from alcoholic biological parents. Archs gen. Psychiat. *28:* 238–243 (1973).
7 Goodwin, D.W.; Schulsinger, F.; Knop, J.; Mednick, S.; Guze, S.B.: Alcoholism and depression in adopted-out daughters of alcoholics. Archs gen. Psychiat. *34:* 751–755 (1977).
8 Little, R.E.; Schultz, F.A.; Mandell, W.: Drinking during pregnancy. J. Stud. Alcohol *37:* 375–379 (1976).
9 MacKay, J.R.: Problem drinking among juvenile delinquents. Crime Delinquency *9:* 29–38 (1963).
10 Roe, A.: The adult adjustment of children of alcoholic parents raised in foster-homes. Q. Jl Stud. Alcohol *5:* 378–393 (1944).
11 Schuckit, M.A.; Goodwin, D.W.; Winokur, G.: A study of alcoholism in half siblings. Am. J. Psychiat. *128:* 122–126 (1972).
12 Sullivan, W.C.: A note on the influence of maternal inebriety on the offspring. J. ment. Sci. *45:* 489–503 (1899).

M.E. Tumbleson, MD, Sinclair Research Farm, University of Missouri, Columbia, MO 65211 (USA)

Voluntary Ethanol Consumption, as a Function of Estrus, in Adult Sinclair(S-1) Miniature Sows[1]

Myron E. Tumbleson, James D. Dexter, Peter Van Cleve

College of Veterinary Medicine, School of Medicine and Sinclair Comparative Medicine Research Farm, University of Missouri, Columbia, Mo., USA

Introduction

Little is known with respect to voluntary ethanol consumption as a function of estrus cycle. During the past few years, we have observed that male pigs have a greater mean voluntary consumption of ethanol than do females. Also, we noted more variability in daily intake by females than by males. Consequently we designed and conducted an experiment to evaluate cyclic voluntary ethanol intake by adult sows.

Materials and Methods

Six 2-year-old Sinclair(S-1) miniature sows were selected for the investigation. Each sow had been allowed access to ethanol for at least 6 months prior to initiation of data collection. Sows were 4–10 weeks postlactation.

Each sow was housed individually in a 1.2 × 2.5 m indoor pen with a concrete floor and chain link dividing fences. Therefore, each pig was able to touch other pigs. Temperature was maintained at 12–15°C in the winter and varied with outside temperature in the summer. Fresh drinking water was supplied ad libitum. Each pig was given a daily ration of 2% of body weight at 08.00 h; in general, the ration was consumed in a 15-min period. Diet composition is depicted in table I.

Individual body weights were recorded biweekly. Daily ad libitum voluntary ethanol consumption was quantitated between 09.00 and 10.00 h. Dietary ethanol was provided, via lixits, as a 20% (w/v) aqueous solution.

---

[1] Supported in part by a grant from the United States Brewers Association.

Table I. Diet composition

Ingredients	kg/100 kg
Soybean meal	40.0
Mean and bone meal	20.0
Alfalfa meal	20.0
Wheat shorts	12.0
Calcium carbonate	1.5
Dicalcium phosphate	3.5
Trace mineral salt	1.0
Vitamin premix	2.0

Table II. Voluntary ethanol consumption by 6 sows during a 364-day period

Sow No.	g ethanol/kg body weight/day				
	mean	<0.25	<1	>6	>8
1156	3.34	17	44	30	8
1172	3.53	56	68	66	23
1163	3.87	14	31	60	6
1173	4.00	26	31	56	7
1157	4.48	16	30	86	32
1188	5.17	26	35	138	56
Mean	4.06	26	40	73	22

Results and Discussion

Mean ethanol consumption for the 6 sows during the year long study was 4.06 g/kg body weight/day (table II). Average total metabolizable calories consumed/pig/day was 3,700 with 39% from ethanol. Mean body weight was 43 kg at the initiation of the study and 59 kg after 12 months on test. Therefore, dietary intakes were adequate for 'normal' growth.

Daily ethanol consumption by each of the 6 sows is presented in figure 1–6. As depicted in the figures and listed in table II, there was considerable variability in consumption patterns and intake/sow/day. Approximately 11% of the days, pigs consumed less than 1 g ethanol/kg body weight/day and 6% of the days, they consumed more than 8 g/kg.

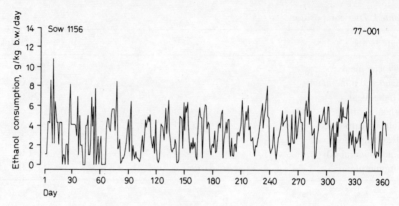

Fig. 1. Daily voluntary ethanol consumption by sow 1156.

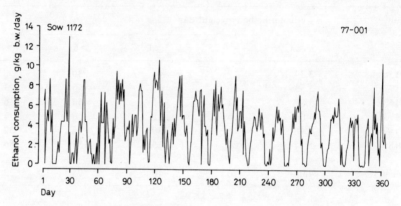

Fig. 2. Daily voluntary ethanol consumption by sow 1172.

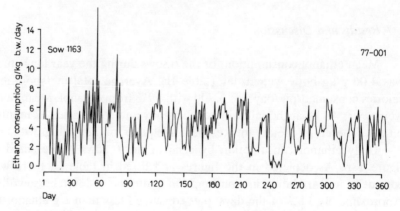

Fig. 3. Daily voluntary ethanol consumption by sow 1163.

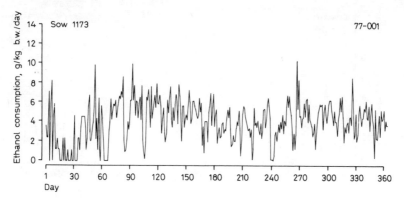

Fig. 4. Daily voluntary ethanol consumption by sow 1173.

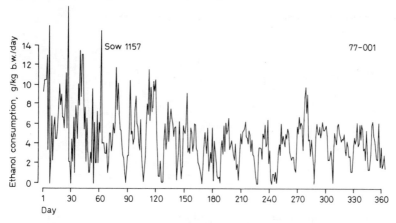

Fig. 5. Daily voluntary ethanol consumption by sow 1157.

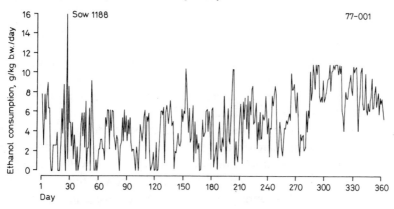

Fig. 6. Daily voluntary ethanol consumption by sow 1188.

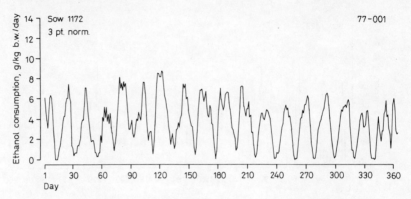

Fig. 7. Three-point normalization of daily ethanol consumption by sow 1172.

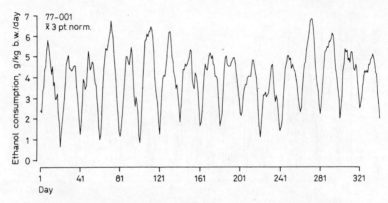

Fig. 8. Three-point normalization of daily ethanol consumption by all 6 sows. Estrus cycle times were matched prior to calculations of means.

For each of the 6 pigs, cyclic intake patterns are more or less apparent at different times during the study. However, when the data for a given animal is 3-point normalized and presented graphically (fig. 7), the cyclic phenomenon is delineated more readily.

Mean estrus cycle duration was 20.2 days. When estrus cycle times for the 6 sows were matched and the mean data 3-point normalized, the cyclic nature of voluntary ethanol consumption is well documented. During estrus, the sows often did not eat the daily ration as quickly as during anestrus. Also, during estrus, the animals were more active and fidgety. Even though there were numerous occasions when the sows consumed negligible quantities of ethanol, they did not appear to experience withdrawal. However, there were a few times when the first indications of withdrawal were noticed. When working with boars, after a day or two of minimal ethanol consumption there often are noticeable withdrawal signs.

During the course of a study of this length, there are days when particular animals appear to be less healthy; therefore, the 3-point data normalization is of assistance in evaluation of patterns. For sow 1163, there was a 2-week period, at about 35 weeks on test, when she experienced a fever and was not in excellent health. Consequently, her consumption was depressed. Also, different special activities in other areas of the barn affected consumption. However, when compiling the data for all 6 sows (fig. 8), it is apparent that voluntary ethanol consumption is affected by stage of estrus.

Summary

Daily voluntary ethanol consumption was determined, for a 1-year period, for 6 adult Sinclair(S-1) miniature sows. Mean ethanol consumptions, for the 6 animals, were 3.34, 3.53, 3.87, 4.00, 4.48 and 5.17 g/kg body weight/day. During the 3-day estrus periods, overall mean intake was 1.68 g/kg body weight/day; whereas, during the 3-day midestrus periods, overall mean ethanol consumption was 5.20 g/kg body weight/day.

M. E. Tumbleson, MD, College of Veterinary Medicine, University of Missouri, Columbia, MO 65211 (USA)

Effects of Hippocampal Lesions on Ethanol Intake in Mice

James N. Pasley, Ervin W. Powell[1]

Departments of Physiology and Anatomy, University of Arkansas for Medical Sciences, Little Rock, Ark., USA

Introduction

Ethanol has differential actions on various parts of the brain and has been postulated to act on 'drinking emotional circuit' components within the limbic system [10]. Structures belonging to the limbic system include the hippocampus, septum and amygdala. The hippocampus is of particular interest because it has been implicated with a number of effects associated with ethanol ingestion [1, 2, 4, 6, 8, 16]. *Hunt and Dalton* [4] demonstrated that hippocampal acetylcholine levels are decreased by chronic ethanol treatment. Moreover, *Goldman* et al. [2] reported that the hippocampus was the only brain area of ten studied that exhibited decreased blood flow after an intoxicating ethanol dose. In addition, hippocampal neurons have been shown to exhibit a high ethanol sensitivity as measured by spontaneous impulse discharge-patterns and evoked response patterns when compared with other areas of the brain [1, 6]. In addition, recent studies have demonstrated morphological alterations in the hippocampus following long-term ethanol consumption [8, 14, 16].

The purpose of the present study, therefore, was to determine ethanol consumption in a free-choice situation after bilateral electrolytic lesions of the hippocampus in order to evaluate the importance of this limbic system structure in the modulation of ethanol ingestion.

---

[1] We thank *W. McGibbony* for excellent technical assistance.

Materials and Methods

Animals and Procedures

The brown house mice *(Mus musculus)* used in this study were from an outbred colony several generations removed from the wild. An initial study was conducted (experiment 1) in which trends in ethanol consumption were observed in various groups of mice with and without lesions in different areas of the hippocampus. We determined volitional selection for ethanol in five groups of 10 adult male mice which had never previously been exposed to alcohol. Surgical operations were performed under anesthesia (Nembutal, Abbott 35 mg/kg, i.p.). The lesions were stereotaxically placed by a direct current lesion maker using coordinates determined by *Lehman* [7]. One group underwent bilateral stereotaxic lesions of the dorsal hippocampus (DH) (A 2.6 L 1.3 H 2.5) (fig. 1a), a second underwent bilateral lesions of the central segment of the hippocampus (CH) (A 1.2 L 2.75 H 2.5) (fig. 1b), a third group underwent lesions of the ventral hippocampus (VH) (A 1.62 L 2.2 H 1.4) (fig. 1c), a fourth group of 10 mice was sham-operated and the fifth group remained intact and unoperated. Sham-lesioning consisted of electrode placement without the passage of current. 1 mA of current passed for 10 s was used to produce a lesion approximately 1 mm in diameter utilizing a stainless steel electrode 350 μm in diameter insulated except for 0.5 mm of the tip.

The mice were singly caged and allowed 6 days recovery after surgery before testing began. Individual preferences for ethanol in the mice were determined by the two-choice, three-bottle technique to avoid position habit [9]. One calibrated drinking tube contained water, another contained a 5% v/v solution of ethanol, and the third bottle remained empty. All drinking tubes were fitted with ball bearings to prevent leakage. Tube positions on the cages were changed daily. All mice were given free access to Purina Laboratory Chow as well as to the two fluids and were maintained on a 12:12 LD cycle at 24 °C. Fluid intakes were recorded at the same time every day (15.00 h). Data were collected for 14 days.

A second study (experiment 2) was subsequently undertaken to investigate further preliminary trends found in the first experiment. 17 additional male mice between 60 and 80 days of age were used in this study. 10 mice of which underwent bilateral lesioning of the CH and 7 were sham-operated (SH). Lesioning procedures were as described above. The ethanol drinking patterns of the mice were determined by the three-bottle, two-choice technique described earlier. The solution of ethanol, however, was increased in concentration of 12 successive days as follows: 3, 4, 5, 6, 7, 9, 11, 13, 15, 20, 25 and 30%. Each volume/volume solution was prepared in tap water with 95% ethanol. This study was undertaken to determine the animals preference threshold which is defined as the concentration at which ethanol constitutes one-half of the animals' daily fluid intake when a series of concentrations are offered in ascending or descending order. Following the 12-day ethanol preference test, the 24-hour drinking patterns of 12 additional control and hippocampal-lesioned mice were examined. Bilateral lesions in the central segment of the hippocampus were then placed in 6 male mice and 6 others served as sham-operated controls. Lesioning and volitional intake measurements were performed as described earlier. The circadian drinking patterns were monitored for 6 days by a multiple-channel event recorder. This study was undertaken to provide data on the circadian ingestive behavior related to ethanol and water intake in mice with and without bilateral lesions of the hippocampus, since bilateral hippocampal lesions have been reported to increase general activity in ro-

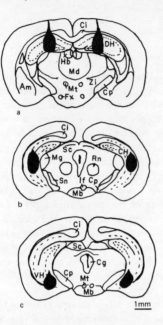

Fig. 1. Diagram of lesion placements in the mouse hippocampus. *a* Example of bilateral lesion placement in the DH. *b* Example of bilateral lesion placement in the CH. *c* Example of bilateral lesion placed in the VH. Am = Amygdala; Cg = cingulate gyrus; CH = central hippocampus; Ci = cingulum; Cp = cerebral peduncle; DH = dorsal hippocampus; Fx = fornix; Hb = habenula; If = interpeduncular fossa; Mb = mammillary body; Md = dorsomedial nucleus of the thalamus; Mg = medial geniculate body; Mt = mammillothalamic tract; Rn = red nucleus; Sc = superior colliculus; Sn = substantia nigra; VH = ventral hippocampus; Zi = zona incerta.

dents [5]. At the end of each study, the mice were killed by cervical dislocation. Brains were removed and fixed in a 10% formalin–30% sucrose solution. Serial frozen sections of brain tissue were cut (30 μm) for localization of electrode and lesion placement [13]. Statistical evaluation of the data included calculation of means and standard errors (SE) and the t test or analysis of variance using the statistical package (Stat Pack) written by the Western Michigan University Computer Center.

Results

The results of the first study are summarized in table I. Average preference levels are expressed as the proportion of ethanol to total fluid intake. The greatest amount of ethanol consumption occurred in CH-lesioned mice. Mice with lesions in the CH selected more ($p<0.01$) of

Table I. The mean proportion of ethanol to total fluid consumed per day for mice (groups of 10 each) with central hippocampal lesions (CH), dorsal hippocampal lesions (DH), ventral hippocampal lesions (VH), sham lesions (SH) and for unoperated (U) mice

Groups of animals	Days													
	1	2	3	4	5	6	7	8	9	10	11	12	13	14
CH	0.53	0.43	0.63	0.51	0.56	0.63	0.66	0.51	0.66	0.58	0.63	0.40	0.65	0.68
DH	0.28	0.36	0.20	0.23	0.05	0.03	0.03	0.03	0.07	0.03	0.01	0.03	0.01	0.05
VH	0.11	0.17	0.09	0.08	0.02	0.04	0.02	0.02	0.09	0.06	0.07	0.02	0.06	0.03
SH	0.02	0.03	0.02	0.04	0.03	0.05	0.02	0.09	0.03	0.02	0.03	0.01	0.02	0.01
U	0.09	0.09	0.04	0.07	0.08	0.09	0.02	0.07	0.02	0.07	0.11	0.05	0.04	0.02

Table II. The proportion of ethanol to total fluid intake and grams of ethanol per kilogram of body weight for sham- (n = 7) and hippocampal-lesioned (n = 10) mice

Concentration of ethanol %	Sham-operated		Hippocampal-lesioned	
	proportion	g/kg	proportion	g/kg
3	0.05 ± 0.02	0.2 ± 0.1	0.46 ± 0.06	2.8 ± 0.4
4	0.03 ± 0.02	0.2 ± 0.1	0.33 ± 0.07	2.7 ± 0.4
5	0.02 ± 0.01	0.1 ± 0.1	0.63 ± 0.30	5.5 ± 1.0
6	0.01 ± 0.01	0.1 ± 0.04	0.61 ± 0.11	6.1 ± 1.3
7	0.01 ± 0.01	0.1 ± 0.1	0.34 ± 0.09	4.0 ± 0.9
9	0.02 ± 0.01	0.3 ± 0.1	0.30 ± 0.08	5.3 ± 1.4
11	0.02 ± 0.01	0.2 ± 0.1	0.16 ± 0.06	2.9 ± 1.0
13	0.01 ± 0.01	0.01 ± 0.1	0.07 ± 0.03	1.6 ± 0.6
15	0.01 ± 0.01	0.2 ± 0.1	0.05 ± 0.01	1.6 ± 0.5
20	0.02 ± 0.01	0.3 ± 0.1	0.05 ± 0.01	2.1 ± 0.5
25	0.01 ± 0.01	0.2 ± 0.1	0.08 ± 0.02	3.1 ± 0.6
30	0.04 ± 0.01	0.1 ± 0.03	0.03 ± 0.01	2.0 ± 0.4

All values are means and their standard errors.

their fluid intake as ethanol over the 14-day period than did the other groups. Those groups of mice with lesions in the DH or VH also selected a substantial proportion of their fluid as ethanol the first 3–4 days. During the remainder of the 14-day preference test, however, these two groups of mice consistently consumed quantities similar to those of controls.

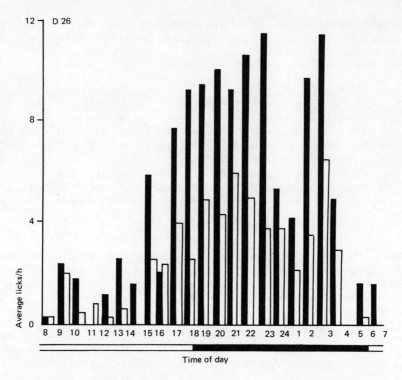

Fig. 2. The 24-hour licking record over 6 days representing water (hatched bars) and 5% ethanol (solid bars) intake in a mouse with a central hippocampal lesion under a 12:12 LD photoperiod.

Results of the preference threshold test of the SH and CH groups are seen in table II. The hippocampal-lesioned mice drank greater amounts of ethanol than controls even though the ethanol concentration was progressively raised from 3 to 30%. As the choice of ethanol concentrations entered the 11–30% range, however, the proportion of ethanol to total fluid intake of CH mice fell substantially. The preference threshold for the CH-lesioned mice was well above that of the SH controls. The CH-lesioned mice consumed a daily average of 6.8 ± 1.1 ml of total fluid and 3.9 ± 0.6 g of food per mouse while the SH mice averaged 6.4 ± 0.5 ml of total fluid and 4.9 ± 0.4 g of food per mouse.

The Drinkometer records indicated that the drinking activity of water and ethanol in both sham-lesioned and hippocampal-lesioned mice

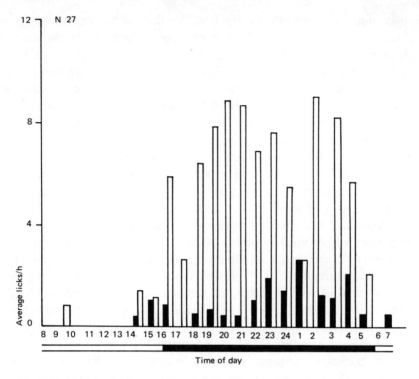

Fig. 3. The 24-hour licking record over 6 days representing water (hatched bars) and 5% ethanol (solid bars) intake in a sham-operated mouse under a 12:12 LD photoperiod.

was pronouncedly circadian. In both groups of mice, most ethanol and water drinking occurred between 17.00 and 04.00 h with maximum intensity most often between 20.00 and 22.00 h. The 24-hour record of licks representing the intake of water and ethanol for the 6-day experimental period is plotted for a representative lesioned mouse in figure 2 and a sham-operated mouse in figure 3.

Discussion

These studies indicate that lesions of the central segment of the hippocampus in the nouse mouse induce drinking of large amounts of ethanol in a free-choice situation with water. Why the DH- and VH-lesioned mice initially drank ethanol the first few days of the experiment is not known. It may, however, have been due to their continuing recovery from surgery or other lesion effects. In addition, the CH-lesioned mice consumed enhanced quantities of ethanol up to a concentration of 11% even though increasing the concentration makes the solution more gustatorily aversive. Lesioning of the hippocampus, however, did not alter 24-hour fluid drinking activity compared with controls.

There may be a factor in the hippocampus which is important in the discrimination-selection systems relating to ethanol ingestion. Although alterations of the biochemical milieu of the brain have been shown to induce changes in ethanol consumption in rats [11], the role of the hippocampus and other neural mechanisms related to fluid selection is not known. The important trace metal, zinc, is highly concentrated in the brain within the hippocampus. The role of zinc in taste and smell activity with accompanying alterations in eating and drinking behavior is well documented [3]. Preliminary studies in our laboratory have shown that chronic ingestion of ethanol results in diminished brain zinc concentrations in mice [12]. Additionally, C57 mice, an ethanol-preferring strain, have markedly lower brain zinc levels compared with their ethanol-nonpreferring counterparts [12]. Perhaps, lesioning of the hippocampus, a brain area rich in zinc, may alter concentrations of this trace metal enough to modify ethanol intake behavior in the mouse.

Thus, the central segment of the hippocampus may play an important role in the appetite for ethanol in the mouse. These findings appear consistent with those from recent reports that long-term ethanol consumption in mice leads to morphological alterations in the hippocampus [8, 14, 16]. In addition, *Siggins and Bloom* [15] have recently reported that the direct application of ethanol in vivo to hippocampal pyramidal cells results in both excitatory and inhibitory actions of spontaneous discharge patterns. Moreover, the reported higher sensitivity of hippocampal neurons to ethanol compared with other brain areas [6] and the depression of evoked responses in the hippocampus by ethanol [1] also appear indirectly supportive of possible hippocampal involvement in ethanol ingestion. Therefore, it seems plausible that ethanol consumption may lead progressively to more and more ethanol intake involving physiological, morphological and topographical features of and effects on the hippocampus, i.e. more ethanol ingestion leading to subsequently more hippocampal damage.

Summary

The effects of bilateral electrolytic hippocampal lesions on voluntary ethanol selection were examined in five groups (n = 10 each) of male mice. Mice sustained lesions in the dorsal hippocampus (DH), central hippocampus (CH), ventral hippocampus (VH), were sham-operated (SH) or left unoperated. CH mice exhibited a preference for ethanol compared with all other groups ($p < 0.01$). In a second experiment, ethanol preference threshold and 24-hour drinking patterns were examined in CH and SH mice. These data revealed a higher preference threshold in CH mice, but no difference in drinking activity periods compared with controls. The central hippocampal area may be important in ethanol ingestive behavior in mice.

References

1   Folger, W.R.; Klemm, W.R.: Unit responsitivity in various hippocampal layers and the depressive effects of ethanol. Physiol. Behav. *21:* 169–176 (1978).
2   Goldman, H.; Sapirstein, L.A.; Murphy, S.; Moore, J.: Alcohol and regional blood flow in brains of rats. Proc. Soc. exp. Biol. Med. *144:* 983–988 (1973).
3   Henkin, R.; Patten, B.; Re, P.; Bronzert, D.: A syndrome of acute zinc loss. Archs Neurol. *32:* 745–751 (1975).
4   Hunt, W.A.; Dalton, T.K.: Regional brain acetylcholine levels in rats acutely treated with ethanol or rendered ethanol-dependent. Brain Res. *109:* 628–631 (1976).
5   Kimble, D.P.: The effects of bilateral hippocampal lesions in rats. J. Comp. physiol. psychol. *56:* 273–282 (1963).
6   Klemm, W.E.; Dreyfus, L.R.; Forney, E.; Mayfield, M.A.: Differential effects of low doses of ethanol on the impulse activity in various regions of the limbic system. Psychopharmacology *50:* 131–138 (1976).
7   Lehman, A.: Atlas stéréotaxique du cerveau de la souris (Editions du Centre National de la Recherche Scientifique, Paris 1974).
8   Montgomery, R.L.; Pick, J.R.; Ellis, F.W.; Christian, E.L.: Microscopic studies in hippocampal and cortical areas of rhesus monkeys physically dependent on ethanol. Anat. Rec. *193:* 627 (1979).
9   Myers, R.D.; Holman, R.B.: A procedure for eliminating position habit in preference-aversion tests for ethanol and other fluids. Psychon. Sci. *6:* 235–236 (1966).
10 Myers, R.D.; Veale, W.L.: The determinants of alcohol preference in animals, in Kissin Begletier, The biology of alcoholism, p. 131 (Plenum, New York 1972).
11 Myers, R.D.; Melchoir, C.L.: Alcohol drinking: abnormal intake caused by tetrahydropapaveroline in brain. Science *196:* 554–556 (1977).
12 Pasley, J.N.; Stull, R.E.; Light, K.E.: Zinc concentrations in two strains of mice after free choice and forced ethanol exposure. Drug Alcohol Depend. *6:* 9–10 (1980).
13 Powell, E.W.: A rapid method for intracranial electrode localization using unstained frozen sections. Electroenceph. clin. Neurophysiol. *17:* 432–434 (1964).
14 Riley, J.N.; Walker, D.W.: Morphological alterations in hippocampus after long-term alcohol consumption in mice. Science *201:* 646–648 (1978).

15 Siggins, G.R.; Bloom, F.E.: Alcohol-related electrophysiology. Drug Alcohol Depend. *6:* 81–82 (1980).
16 Walker, D.W.; Barnes, D.E.; Zornetzer, S.F.; Hunter, B.E.; Kubanis, P.: Neuronal loss in hippocampus induced by prolonged ethanol consumption in rats. Science *209:* 711–713 (1980).

Dr. James N. Pasley, Department of Physiology, University of Arkansas,
Little Rock, AR 72205 (USA)

Steroidal Actions and Voluntary Drinking of Ethanol by Male and Female Rats[1]

F.S. Messiha[2]

Division of Toxicology, Department of Pathology, and Psychopharmacology Laboratory, Department of Psychiatry, Texas Tech University Health Sciences Center, School of Medicine, Lubbock, Tex., USA

Introduction

An adequate animal model for voluntary ethanol (ET) consumption has yet to be developed. Nevertheless, the high metabolic rate of rodents toward biotransformation of various drugs, compared to man, and the ability of the rat to select a diluted ET solution over water as the drinking fluid of choice [20] provides one of the experimental models available. This behavioral performance model has often been used to study the effect of pharmacological interventions on voluntary intake of ET and to evaluate ET-drug interactions in conjunction with probing into the possible underlying mechanism(s). The well-known abnormal endocrine function in chronic alcoholism has provided the experimental rational for the use of various steroidal agents in intact and in gonadectomized rats with preference to ET [7, 12, 21, 24, 26]. The present study reports on the effect of certain steroidal agents with estrogenic and antiestrogenic properties and of a nonsteroidal antiandrogen on voluntary intake of ET by the rat. In addition, determinations of hepatic alcohol dehydrogenase (L-ADH) and aldehyde dehydrogenase (L-ALDH) were made in these animals after drug administration.

[1] Supported in part by a grant from TTUHSC Biomedical Research Institute.
[2] The competent and excellent technical assistance of Mr. *James Webb* is acknowledged. Thanks are due to Drs. *Rudolph Neri* and *David H. McCurdy* for their generous supplies of flutamide and SCH-16423 and for tamoxifen, respectively.

Materials and Methods

Sexually mature male and female rats of the Sprague-Dawley strain (Holtzman Farm Co., Madison, Wisc.) were used throughout. The animals were 70 days old at the beginning of the experiments. The female rats were housed in complete darkness for 7 consecutive days to minimize marked changes of their estrus cycle, and were then housed in a room with 12 h dark and 12 h light cycles as the male rats. They were provided with Purina pellet chow and water ad libitum for 10 days prior to the initiation of the behavioral experiments.

Behavioral Experiments

Animals were offered 5 or 10% ET (w/w) solution, prepared from 95% ETOH, as the only drinking fluid (habituation period) prior to exposing them to free choice between water and the ET solution for 3 weeks as described in detail elsewhere [14]. In experiments with male rats, only those who consumed over 60% of their daily intake of fluid as ET for 3 weeks under the free-choice condition were used. This is compared to the selection of experimental female rats who consumed at least 30% of the ET solution as the drinking fluid daily for an identical duration of time as the male rats. Fluid intake and food consumptions were measured daily, approximately 4 h after the beginning of the light cycle, 11.00 a.m.

In the first set of experiments, groups of male and female rats with preference to ET drinking, on the basis cited above, were administered estradiol, 0.3 mg/kg i.p. or ethinyl estradiol, 0.3 mg/kg i.p., respectively. The animals were sacrificed 24 h after drug administration, their fluid and food consuptions were measured, and livers were used for the enzymatic determinations of L-ADH and L-ALDH as mentioned below. The controls consisted of a separate group of animals of both sexes who preferred 5% ET drinking and received the vehicle, vegetable oil, intraperitoneally and were sacrificed for the hepatic enzymatic assays. The changes in water, ET and food consumptions for the 24 h after drug administration are compared with the mean corresponding values obtained for the same animals for the 24 h prior to drug treatment and are also compared with that obtained for the preceding 4 consecutive days.

In the second set of experiments, the effect of antiestrogenic steroid tamoxifen citrate (TMX) on the voluntary drinking of 10% ET solution was tested in both male and female rats. The drug was administered once, 1.0 mg/kg p.o., and the drinking pattern of water and ET was followed longitudinally. 9 days thereafter TMX was given, 2 mg/kg p.o., once daily for 5 consecutive days prior to sacrifice of the animals for the measurements of liver enzymes. The controls consisted of animals with preference to ET drinking who received the vehicle, vegetable oil, at identical time intervals as the drug-treatment group.

In the third set of experiments the effects of the nonsteroidal compound antiandrogen flutamide (FL) and its metabolite SCH-16423 on voluntary selection of ET were studied in male rats. These compounds were administered, 20 mg/kg p.o., once daily for 2 consecutive days and measurements of the daily intake of drinking fluids and food consumption were recorded.

Drugs were of analytical grade (Sigma Chemical Co). They were dissolved in organic solvent, suspended in vegetable oil and the organic solvent evaporated over water bath. The vehicle used for the control rats was prepared by suspending an identical volume of the organic solvent into the vegetable oil and evaporating of the solvent in a similar fashion.

Biochemical Experiments

Livers were quickly removed from the sacrificed animals and washed with 0.1 M KCl, pH 6.8, homogenized in 5 vol of the KCl buffer with a Waring blender. The homogenates were subjected to differential centrifugation to obtain the mitochondrial (M) and the cytoplasmic (C) fractions as described in detail elsewhere [15]. Cytosolic ALDH [3] and ADH [2] were assayed by established procedures. The mitochondrial ALDH enzymes with the apparent low K_m and the apparent high K_m value were also determined [22]. Portions of each fraction served for the protein assay by the biuret method. The enzymatic activity is expressed as specific activity, nmol/min/mg protein, measured at 30 °C.

The results are means ± SE and the statistical significance of the data was analyzed by two tailed Student's t test for independent means.

Results

Figure 1 shows the effect of administration of an identical single dose, 0.3 mg/kg i.p., of a naturally occurring estrogen, i.e. estradiol, and a synthetic compound ethinyl estradiol on voluntary intake of the 5% ET solution by the male and female rat, respectively. Injection of estradiol to the male rat (fig. 1a) decreased voluntary drinking of the 5% ET solution by approximately 56% ($p<0.02$) and 48% ($p<0.01$) from the previous 24 h or from the mean daily intake for the preceding 4 days, respectively. This decrease in ET intake was accompanied by a marked rise in water intake with little changes occurring in food consumption. Injection of the synthetic estrogen, ethinyl estradiol, 0.3 mg/kg i.p., to female rats exerted a profound reduction on voluntary drinking of ET. Ethinyl estradiol decreased the selection of ET for drinking by approximately 73% ($p<0.005$) from the mean consumption during an estrus cycle and this decrease was also evident when compared to the amount of ET consumed for the 24-hour period before drug treatment ($p<0.05$). Increased water consumption following the ethinyl estradiol injection was also noted ($p<0.05$).

Figure 2 illustrates specific activities of rat liver enzymes studied as a function of estrogenic treatment, 0.3 mg/kg i.p. Administration of the synthetic estrogenic drug, ethinyl estradiol, to females with preference to ET exerted little effect on the enzymes measured from corresponding control female rats with preference to the 5% ET solution but injected the vehicle. In male rats preferring ET, the administration of estradiol produced minimal changes in specific activities of L-ADH and L-ALDH in the cytosolic fraction compared to respective controls. Estradiol administration decreased the total mitochondrial L-ALDH activity by approximately 16% from controls. However, this decrease was not statistically

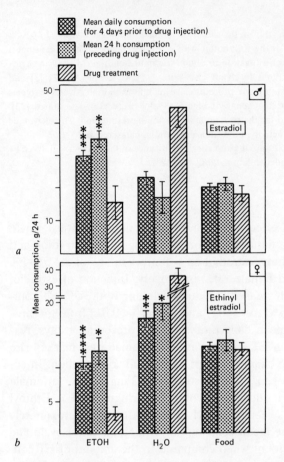

Fig. 1. The effect of identical intraperitoneal doses of estradiol and ethinyl estradiol, 0.3 mg/kg, on voluntary intake of 5% (w/w) ethanol (ETOH) solution, water and food consumptions by the male (a) and by female (b) rat. Each bar graph represents mean ± SE of 24-hour consumptions of water, 5% ETOH and food for 5–6 rats. * $p < 0.05$; ** $p < 0.02$; *** $p < 0.01$; **** $p < 0.005$.

significant. Differential estimation of L-M-ALDH into its two components indicates that estradiol-produced inhibition was confined to the mitochondrial enzyme with the apparent high K_m value ($p < 0.05$).

Figure 3 shows the effect of a single and multiple dose administration of an antiestrogenic steroidal compound, TMX, on female and male rats drinking the 10% (w/w) ET solution in the presence of water under

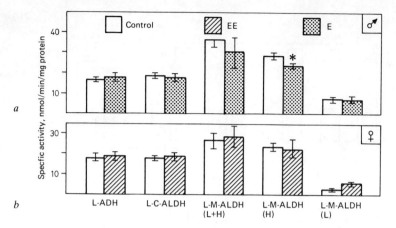

Fig. 2. The effect of administration of estradiol (E), 0.2 mg/kg i.p., to male rats (a) and ethinyl estradiol (EE), 0.3 mg/kg i.p., to female rats (b) with preference to ethanol (ET) on endogenous hepatic ET-metabolizing enzymes. Animals were sacrificed 24 h after drug administration and liver alcohol dehydrogenase (L-ADH) and aldehyde dehydrogenase (L-ALDH) were assayed in the cytosolic (C) fraction. Mitochondrial (M) L-ALDH was measured for the enzyme with the apparent low (L) and high (H) K_m value and their combination (L+H). Each bar graph represents means ± SE of at least 6 independent determinations. * $p<0.05$.

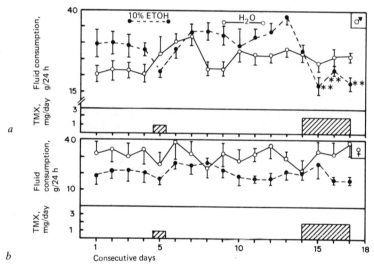

Fig. 3. The effect of oral administration of an antiestrogenic compound, tamoxifen citrate (TMX) on the behavioral voluntary drinking profile of 10% (w/w) ethanol (ETOH) by the male (a) and female (b) rat with preference to ETOH. Each point represents mean ± SE of daily fluid consumption (g/24 h) for 5 animals. TMX was administered in the dose and for the duration of time indicated in the lower panel of each graph. ** $p<0.02$.

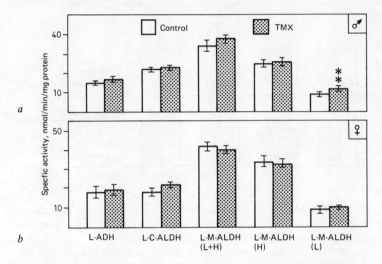

Fig. 4. The effect of oral administration of tamoxifen citrate (TMX), 2 mg/kg/day for 5 consecutive days, on hepatic ethanol and acetaldehyde-metabolizing enzymes in the male (a) and female (b) rat with preference to ethanol. Animals with preference to 10% ethanol solution were sacrificed 1 h after the final drug or vehicle injection. Liver alcohol dehydrogenase (L-ADH) and aldehyde dehydrogenase (L-ALDH) were determined in the cytosolic (C) fraction. Measurements of L-ALDH were also made in the mitochondrial (M) fraction for the enzyme with the apparent low (L) and high (H) K_m value and their combination (L+H). Each bar graph represents mean ± SE of specific activity (nmol/min/mg protein) derived from 5 animals.

the free-choice situation. Oral administration of TMX, 1 mg/kg, decreased voluntary ET drinking by approximately 25% in both male ($p<0.1$) and female rats. TMX was readministered 8 days after the initial dose to the same animals. The dose regimens consisted of the administration of a greater dose of TMX, 2 mg/kg p.o., once daily for 5 consecutive days prior to sacrifice of the animals for the biochemical studies. There was a 28% ($p<0.1$) decrease in ET consumption by the male rats after the initial 2 mg/kg dose of TMX. Continued administration of TMX resulted in a further decrease in voluntary drinking of ET after the second TMX dose ($p<0.02$). This significant decrease persisted during the subsequent administration of TMX ($p<0.02$). Mean intake of ET consumption during TMX treatment was 21.5 ± 2.3 compared to 31.2 ± 2.2 g/24 h mean daily intake for 5 days preceding drug administration to the male rat ($p<0.02$). Administration of identical dose regimens of TMX to the female rat did not produce aversion to ET drinking.

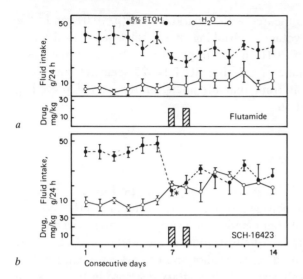

Fig. 5. The effect of antiandrogen on voluntary drinking of 5% (w/w) ethanol (ETOH) solution over water by the male rat. Flutamide and its hydroxylated metabolite SCH-16423 were injected, 20 mg/kg i.p., once daily for 2 consecutive days. Each point represents mean ± SE of daily ETOH, water and food consumption (g/24 h) for 5 adult male rats for each drug treatment.

The effect of short-term administration of TMX, 2 mg/kg, p.o./day for 5 days on hepatic ET-metabolizing enzymes of the animals used for the behavioral experiment (fig. 3) is presented in figure 4. Oral administration of TMX to male rats did not alter specific activity of L-ADH or L-ALDH in the cytoplasma compared to the controls receiving the vehicle and exhibiting similar voluntary ET-drinking behavior as the treatment group. There was a moderate but significant reduction occurring in hepatic mitochondrial ALDH enzyme with the apparent low K_m value ($p<0.02$). The dose schedule used of TMX in the female rat was not associated with changes in specific activities of L-ADH and L-ALDH in all fractions measured from respective controls.

Figure 5 shows the effect of the antiandrogen FL and its hydroxylated biologically active metabolite (SCH-16423) on voluntary drinking of 5% ET by the male rat. Intraperitoneal injection of these agents, 20 mg/kg, once daily for 2 consecutive days produced a variable degree of alteration of ET preference. For example, administration of FL showed a slight but not significant decrease in voluntary consumption of

ET by the rat. This is compared with a marked 66% ($p<0.01$) and 55% ($p<0.01$) decrease in voluntary drinking of ET subsequent to the first and the second dose of SCH-16423, respectively. Moreover, the baseline for the average intake of ET after SCH-16423 seems to have been reduced to some extent from that obtained for the period preceding SCH-16423 administration.

Discussion

The present study utilizes an experimental performance test to assess the relationships between voluntary ET drinking, differences in sex and certain steroidal compounds with estrogenic, antiestrogenic, and antiandrogenic properties. Administration of an identical dose of the estrogenic compounds used, i.e. estradiol and ethinyl estradiol, resulted in aversion to ET in both sexes. The voluntary decrease in ET intake in the male rat after injection of estradiol is consistent with that reported in deer mice [5] and in the castrated male rat [1]. Likewise, the ethinyl estradiol-produced reduction in the ET-drinking behavior of the female rat is in agreement with that reported for other synthetic estrogenic compounds of oral contraceptives of the estrogenic-progestine type [6, 13]. Although it is difficult to extend experimental findings to clinical observations, nonetheless the foregoing observations suggest that estrogenic compounds are involved in mediating aversion to ET drinking in subjects with preference to ET. A similar inference to estrogen-induced abnormal endocrine response in alcoholism can be seen from reports indicating disturbed estradiol clearance protein binding associated with demasculinization of some alcoholic men [8]; disturbance in menstrual cycle [11]; voluntary decreased consumption of alcoholic beverages during first trimester and during pregnancy in alcoholic women; after conception of monkeys [4] and miniature swine [23], and during the estrous phase of the cycle at the time of elevated derangements of circulating estradiol levels [11]. The mechanism by which estrogens mediate their actions on ET preference has rarely been studied. However, it is conceivable that the observed inhibition of hepatic mitochondrial ALDH by a single dose of estradiol or by short-term administration of ethinyl estradiol to female rats [16] may underly the mechanism by which an estrogenic steroidal compound may negate ET preference. This inhibition L-M-ALDH results in accumulation of ET-derived acetaldehyde and causes a disulfiram-like toxic manifestation which produces the aversion to ET drinking.

The lack of changes in L-M-ALDH following a single-dose injection of ethinyl estradiol compared to that noted after short-term administration of ethinyl estradiol to female rats [17] indicates a greater potency for naturally occurring estradiol than that of the synthetic steroid ethinyl estradiol and may also suggest a possible competition between endogenous estradiol with the exogenously administered ethinyl estradiol towards L-M-ALDH. Moreover, it should be noted that female subjects tend to show higher blood plasma ET concentration than men receiving an identical dose [9] and that also women on oral contraceptives containing estrogens metabolize exogenously administered ET slower than those without medication [10]. These reports, along with the present and previous [18] experimental data related to inhibition of the hepatic acetaldehyde-metabolizing enzyme, provide one of possible mechanisms of action related to estrogens and aversion to ET drinking. This in turn may indicate the existence of endogenous mechanisms to alleviate from the toxicity of ET-derived acetaldehyde on the fetus.

In the present study short-term administration of antiestrogenic compound TMX profoundly reduced voluntary intake of ET by the male and not by the female rat. This coincided with inhibition of L-M-ALDH enzyme with the low K_m value, which is also known to be inhibited by disulfiram, by TMX treatment in the male but not in the female rat. This inhibition is at least one of the probable causes for aversion to ET drinking by the male rat. This result is also indicative of sex differences in the action of TMX on this enzyme. Conversely, administration of the nonsteroidal antiandrogen tested or its metabolite were devoid of biochemical effects on these hepatic enzymes [16] but the metabolite was more potent than the parent compound in reducing voluntary drinking of ET by the rat. This observed effect indicates that SCH-16423 is biologically more active than the parent substance. Moreover, the antiandrogenic action of this compound may be related in some way to the behavioral response studied since sexual maturity in the male rat is associated with a decreased metabolic rate of ET, and gonadectomy negates this reduction subsequent to treatment with testosterone which in turn replenishes the loss in activity of L-ADH [19].

The data presented and that available in the literature suggest a hepatic-gonadal link in aversion to ET drinking under free-choice situations which may be mediated by estrogenic and antiandrogenic mechanisms. The results also indicate the inhibitory action of compounds with steroidal configuration on hepatic ALDH as contrasted with the utilization of certain other steroids by L-ADH as substrates [25].

Summary

The effect of certain steroidal compounds with estrogenic and antiestrogenic properties on voluntary intake of ET by rats with preference to ET of both sexes was studied. The effect of a nonsteroidal antiandrogenic compound and its biologically active metabolite (SCH-16423) on these parameters was also evaluated. The observed behavioral drinking profile of ET was studied in conjunction with determination of L-ADH and L-ALDH as a function of drug treatment in both sexes. Acute administration of estrogenic compounds, estradiol and ethinyl estradiol, decreased voluntary drinking of ET by male and female rats, respectively. Short-term administration of the antiestrogenic compound TMX resulted in aversion to ET consumption in the male but not in the female rat. Acute administration of the SCH-16423 but not its parent nonsteroidal antiandrogen compound FL decreased ET drinking from predrug treatment. The biochemical results show an inhibitory action by the steroidal compound on hepatic L-M-ALDH with a lack of biochemical effect of FL and SCH-16423 on these enzymes. The results are discussed in conjunction with those available in literature to propose a hepatic-gonadal link in the preference to ET drinking and indicate that the underlying mechanism is probably related to the inhibitory action of estrogenic compounds used on L-M-ALDH.

References

1. Aschkenasy-Lelu, P.: Relation entre L'effet inhibiteur des œstrogenes sur la consumation spontanée d'alcool par le rat. Rev. fr. Etudes. clin. Biol. *5:* 132–138 (1960).
2. Blair, A. H.; Valèe, B. L.: Some catalytic properties of human liver alcohol dehydrogenase. Biochemistry, N.Y. *5:* 2026–2034 (1966).
3. Blair, A. H.; Bodly, F. H.: Human liver aldehyde dehydrogenase: partial purification and properties. Can. J. Biochem. *47:* 265–272 (1969).
4. Elton, R.; Wilson, E.: Changes in ethanol consumption by pregnant pig-tailed macaques. J. Stud. Alcohol *38:* 2181–2183 (1977).
5. Emerson, G. A.; Brown, R. G.; Nash, T. B.; Moore, W. T.: Species variations in preference for alcohol and in effects of diet or drugs on this preference. J. Pharmac. exp. Ther. *106:* 384–389 (1952).
6. Erickson, K.: Effects of ovarectomy and contraceptive hormones on voluntary alcohol consumption in the albino rat. Jap. J. Stud. Alcohol, *4:* 1–5 (1969).
7. Gerdes, H.: Alcohol und Endokrinum. Internist *19:* 89–96 (1978).
8. Gordon, G. G.; Altman, A.; Southeren, A.; Rubin, E.; Lieber, C.: Effect of alcohol administration on sex hormone metabolism in normal men. New Engl. J. Med. *295:* 793–797 (1976).

9   Jones, B.M.; Jones, M.K.: Alcohol effects in women during menstrual cycle. Ann. N.Y. Acad. Sci. *273:* 576–587 (1976).
10 Jones, B.M.; Jones, M.K.; Pardes, A.: Oral contraceptives and ethanol metabolism. Alcohol Tech. Rep. *5:* 28–32 (1976).
11 Little, R.E.: Drinking during pregnancy: implication for public health. Alcohol Health Res. World *4:* 36–42 (1979).
12 Lloyd, C.W.; Williams, R.H.: Endocrine changes associated with Laennec's cirrhosis of the liver. Am. J. Med. *4:* 315–330 (1948).
13 Mardones, J.: Experimentally induced changes in the free selection of ethanol. Int. Rev. Neurobiol. *100:* 41–76 (1960).
14 Messiha, F.S.: Voluntary drinking of ethanol by the rat: Biogenic amines and possible underlying mechanism. Pharmacol. Biochem. Behav. *9:* 379–384 (1978).
15 Messiha, F.S.; Hughes, M.J.: Liver alcohol and aldehyde dehydrogenase: inhibition and potentiation by histamine agonists and antagonists. Clin. exp. Pharmacol. Physiol. *6:* 281–292 (1979).
16 Messiha, F.S.: Androgens, antiandrogens and voluntary intake of ethanol by the male rat. Res. Commun. Subs. Abuse *1:* 1–8 (1980).
17 Messiha, F.S.; Lox, C.D.; Heine, W.: Studies on ethanol and oral contraceptives: feasibility of a hepatic-gonadal link. Res. Commun. Subs. Abuse *1:* 315–333 (1980).
18 Messiha, F.S.; Lox, C.D.; Sproat, H.F.: Effect of castration and oral contraceptives on hepatic ethanol and acetaldehyde metabolizing enzymes in the male rat. Substance Alcohol Actions/Misuse *1:* 197–202 (1980).
19 Rachamin, G.; MacDonald, A.; Wahid, S.; Clapp, J.; Kahanna, J.; Israel, Y.: Modulation of alcohol dehydrogenase and ethanol metabolism by sex hormones in the spontaneously hypertensive rat. Biochem. J. *186:* 483–490 (1980).
20 Richter, C.P.; Campbell, K.H.: Taste thresholds and taste preference of rats for 5 common sugars. J. Nutr. *20:* 31–46 (1940).
21 Simanowsky, N.P.: Über die Gesundheits-Schädlichkeit hefetrüber Biere und über den Ablauf der künstlichen Verdauung bei Bierzusatz. Arch. Hyg., Berl. *4:* 1–6 (1886).
22 Tottmar, O.; Marchner, H.: Disulfiram as a tool in the studies on the metabolism of acetaldehyde in rats. Acta pharmac. tox. *38:* 366–375 (1976).
23 Tumbleson, M.E.; Cleve, P. van; Dexter, J.D.; Tinsley, J.L.; Middleton, C.C.: Voluntary ethanol consumption as a function of estrus in adult, Sinclair (S-1) miniature sows. Pharmacol. Biochem. Behav. *12:* 325 (1980).
24 Thiel, D.H. van; Lester, R.: The effect of chronic alcohol abuse on sexual function. Clin. Endocrinol. Metab. *8:* 499–510 (1979).
25 Waller, G.; Theorell, H.: Liver alcohol dehydrogenase as a 3-β-hydroxy-5β-cholamic acid dehydrogenase. Archs. Biochem. Biophys. pp. 671–684 (1965).
26 Wright, J.: Endocrine effects of alcohol. Clin. Endocrinol. Metab. *7:* 351–373 (1978).

Dr. F.S. Messiha, Department of Pathology, Texas Tech University,
Health Sciences Center, School of Medicine, Lubbock, TX 79430 (USA)

Amino Acids

Effect of Ethanol upon Placental Uptake of Amino Acids[1]

Stanley E. Fisher[2], Mark Atkinson, Ian Holzman, Ronald David, David H. Van Thiel

Children's Hospital of Pittsburgh and the University of Pittsburgh School of Medicine, Pittsburgh, Pa., USA

Introduction

Alcohol abuse is a major health problem worldwide [1]. The consequences of chronic maternal ethanol abuse are manifested in children as the fetal alcohol syndrome (FAS) [5]. The pathophysiology of the FAS is not completely understood. In general, it is considered to be related primarily to direct toxic effects of ethanol or one of its metabolites upon fetal organ development [5–7, 17]. Maternal nutrition has not been thought to be contributory, especially since maternal malnutrition per se does not produce the FAS [13]. On the other hand, certain features of the FAS, particularly microcephaly and low birth weight and length, suggest that the fetus may have experienced in utero nutrient deprivation, i.e. selective fetal malnutrition. Many animal models, utilized to study the pathogenesis of the FAS, have controlled for maternal caloric intake [2, 4, 7, 16, 17]. However, such studies have not examined the possibility that maternal to fetal transport of essential nutrients may have been impaired, in the face of maternal nutritional sufficiency. To evaluate that question, we have investigated the effect of ethanol upon both in vivo and in vitro placental uptake of amino acids.

---

[1] This manuscript is dedicated to the memory of *Richard Weitzman*, MD (SEF). The authors wish to thank Mrs. *Charlotte Grayson* for her help in obtaining the human placenta.

[2] Supported in part by grants from the National Council on Alcoholism, and a grant from the Hunt Foundation, the Mary H. and W. Clark Hagen Foundation and the NIH (AA04425).

Materials and Methods

In vivo Chronically Catheterized Pregnant Sheep. Pregnant mixed breed ewes were operated upon under spinal anesthesia at 115–120 days of gestation (term = 150 days). Polyethylene catheters were placed in the maternal femoral artery and vein. Following laparotomy and hysterotomy, catheters were inserted in a large uterine vein, the umbilical vein and the fetal descending aorta. All catheters were brought out via the maternal flank for subsequent ethanol infusion or blood sampling.

Following 4 days of recovery, during which fetal acid/base status was monitored and found to be normal, 25 ml of 10% ethanol/kg ewe weight was infused into the ewe via the femoral vein over 60 min. Blood samples were obtained before and at 30-min intervals after initiation of the infusion. Blood amino acid concentrations were determined on a Beckman 121 amino acid analyzer equipped with a Honeywell recorder. Oxygen content, an indicator of blood flow, was measured on a Lex-O_2-Con (Lexington Instrument, Waltham, Mass.). Blood pH and glucose concentrations were determined for both ewe and fetus by standard clinical laboratory methods.

In vitro Human Placental Villi. Term placentas were placed in ice-cold saline immediately upon delivery and all experiments were begun within 30 min. Radiolabeled carboxyl inulin (^{14}C-In) (2.0 mCi/g, >99% pure), ^{14}C alpha-amino isobutyric acid (^{14}C-AIB) 19.9 Ci/mol, >99% pure) and NCS solubilizer were obtained from New England Nuclear (Boston, Mass.); scintillation fluid (PCS) from Amersham Searle (Arlington Heights, Ill.); punctilious ethanol (95%) from Mallinckrodt (St. Louis, Mo.); acetaldehyde (99% pure) from Arthur H. Thomas (Philadelphia, Pa.); Earle's medium (balanced salt plus 100 mg glucose/dl, pH = 7.4) from Microbiological Associates (Walkersville, Md.); nonradioactive AIB from Sigma (St. Louis, Mo.).

Human placental extracellular water (^{14}C-In space) and villus uptake of ^{14}C-AIB were determined by a modification of the methods of *Smith* et al. [18]. Specifically, preincubation in sealed flasks was for 45 min at 37 °C in 14.5 ml of medium, which had been gassed for 3 min with 95% O_2/5% CO_2. Tissue was then incubated in 5.5 ml of fresh, gassed medium in sealed flasks, containing ^{14}C-AIB, with the final concentration of AIB adjusted to 0.5 mM. For ethanol exposed tissue, ethanol (300 mg/dl) was added to both preincubation and incubation media. In contrast, for acetaldehyde experiments, acetaldehyde was not in the preincubation medium. Instead, acetaldehyde final concentration 0.05, 0.2, 2, 5, 10 or 20 mM) was added by injection into sealed flasks for ^{14}C-AIB incubation containing fresh, gassed medium. For evaluation of residual effects of acetaldehyde, the routine 45-min preincubation was followed by a second preincubation of 60 min with fresh medium containing 10 mM acetaldehyde. Control tissue also received an additional preincubation. Following the second preincubation, tissue was washed 3 times with 5 ml of medium, placed in 5.5 ml medium and gassed for 3 min before sealing and incubating with ^{14}C-AIB for 30–90 min.

Tissue content of ^{14}C-AIB or ^{14}C-In was determined after dissolving the pressed tissue (500 g weight) in NCS at 60 °C overnight. Uptake of AIB is expressed as the ratio of the intracellular concentration of ^{14}C-AIB to the extracellular concentration (C_i/C_o) [23]. Data are expressed as the mean ± SEM. Statistical significance was determined by Student's t test for paired data, two-tail, and by analysis of variance. Results are considered probably significant for $p<0.05$ and significant for $p<0.01$.

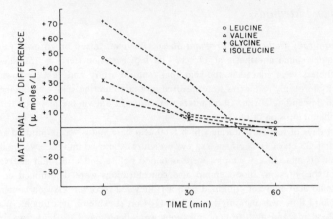

Fig. 1. Effect of ethanol infusion upon placental uptake of amino acids in the pregnant ewe. The fall in maternal uterine A-V difference indicates inhibition of placental uptake. This graph is representative of experiments in 3 sheep.

Results

In vivo. In the pregnant ewe, ethanol infusion produced blood alcohol levels of 150–190 mg/dl by 1 h and 190–260 mg/dl by 2 h. Uterine blood O_2 content was unchanged while fetal and maternal pH and glucose remained normal. Placental uptake of amino acids fell rapidly during ethanol infusion. This is illustrated by the maternal uterine arteriovenous (A-V) difference in blood amino acid concentration (fig. 1). Prior to ethanol infusion, the A-V difference was positive, being as high as +70 µM, (for glycine), which indicates placental uptake. During ethanol exposure, the A-V difference fell to or below 0.

In vitro. In direct contrast to the in vivo findings, ethanol (300 mg/dl) did not affect in vitro human placental villous uptake of AIB (fig. 2). Similarly, 0.05 and 0.2 mM acetaldehyde had no effect upon AIB uptake. Acetaldehyde at pharmacologic concentrations of 2, 5, 10, and 20 mM showed a significant inhibitory effect upon AIB uptake (fig. 3). Comparing control tissue to tissue exposed to acetaldehyde, % inhibition at 90 min was 23.5 ± 9.6% for 2 mM; 33.8 ± 4.7% for 5 mM; 54.6 ± 2.9% for 10 mM; 61.5 ± 7.5% for 20 mM ($p < 0.01$ at all 4 concentrations). When placental villi were exposed to 10 mM acetaldehyde for 60 min of

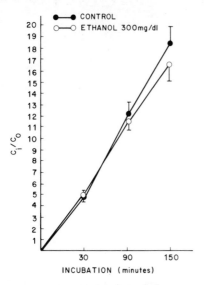

Fig. 2. Effect of ethanol upon AIB uptake by in vitro human placental villi. Incubation time represents the time of ^{14}C-AIB inclusion in the medium. Ethanol exposure (300 mg/dl) began 45 min earlier, during preincubation, e.g. 150 min incubation = 195 min ethanol exposure. There is no significant difference at 30, 90 or 150 min incubation (control and ethanol, n = 16 each).

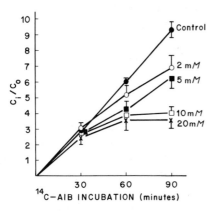

Fig. 3. Effect of acetaldehyde upon AIB uptake by in vitro human placental villi. There was no difference from controls (n = 34) at 30 min. After 60 min of incubation, 5, 10 and 20 mM acetaldehyde significantly reduced ^{14}C-AIB uptake ($p<0.01$). Uptake was reduced for all 4 concentrations at 90 min ($p<0.01$). n = 8 for 2, 10 and 20 mM; n = 10 for 5 mM.

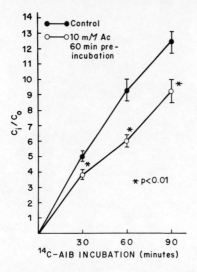

Fig. 4. Residual effect of exposure to 10 μM acetaldehyde upon placental uptake of AIB. Following 30 min of preincubation with acetaldehyde, uptake of ^{14}C-AIB (in the absence of acetaldehyde) was significantly reduced as compared to controls (n = 8) at 30, 60 and 90 min (n = 8 for each time).

preincubation and then incubated with ^{14}C-AIB in the absence of acetaldehyde, significant reduction in AIB uptake occurred at 30, 60 and 90 min (fig. 4). The effect was partially reversible, however, with progressive, linear uptake over time.

Discussion

Under acute in vivo conditions, ethanol appears to rapidly block the transport of amino acids across the ovine placenta. This would be in agreement with rodent studies using both acute and chronic administration of ethanol [8]. Under in vitro conditions, however, physiologic concentrations of ethanol or acetaldehyde failed to alter ^{14}C-AIB uptake by the human placenta. In the chronic alcoholic human, blood ethanol levels usually do not exceed 300 mg/dl and acetaldehyde, the major circulating metabolite of ethanol, usually does not exceed 50 μM [9, 14]. The in vitro system has the advantage of measuring placental concentration of an actively transported, nonmetabolized amino acid, which shares transport mechanisms with most neutral amino acids. The in vitro system is independent of alterations in blood flow, hypoxia or acidosis, which might occur in vivo. It would appear, therefore, that the human placenta is relatively resistant to ethanol-associated effects upon amino acid transport in vitro.

However, under in vivo conditions, especially chronic, ethanol or acetaldehyde could still be toxic to the human placenta, as with the sheep and rodent [8, 20]. Chronic maternal ethanol ingestion might alter placental villus membrane structure or sodium-dependent ATP-ase activity [19]. Both of these factors are important to placental amino acid transport, a membrane-associated, energy-dependent process [15]. Furthermore, amino acid transport is thought to require interaction with protein carriers in the membrane [19]. Ethanol and/or acetaldehyde could alter villus cell synthesis of membrane-associated proteins [20]. Inhibition of protein synthesis might explain the lack of acetaldehyde effect upon AIB uptake at 30 min of incubation. By 60 min, 10 mM acetaldehyde had shut off AIB uptake during continuous exposure. While this might have been the result of cell death, the partial reversibility suggests that cellular function was impaired but simple cell death was not responsible. Nonetheless, even if chronic maternal ethanol ingestion merely 'poisoned' placental cells, leading to cell death, this would be important since placental villus cells cannot regenerate.

Membrane-associated active transport of amino acids by the placenta is similar to that in the intestine, both organs having a villous structure and specialized absorptive cells. Ethanol has been shown to impair amino acid absorption in the animal and human intestine [19]. As with our in vitro findings in the human placenta, high concentrations of ethanol (greater than 1 g/dl) seem to be required to alter intestinal amino acid uptake [3]. Like the intestine, the placenta has the capacity to oxidize ethanol to acetaldehyde [11]. The human placenta is also capable of metabolizing acetaldehyde, although the human placenta contains considerably less capacity to oxidize acetaldehyde than does the rodent placenta [10]. During in vivo exposure, especially chronic, tissue levels of ethanol or acetaldehyde could become sufficiently elevated to alter human placental villus cell function. In this manner, the discrepancy between our in vitro human and in vivo ovine placental data might be explained.

There is little doubt that acute or chronic maternal ethanol exposure is directly harmful to the fetus, at both a morphologic and biochemical level [2, 4, 6, 8, 12, 16, 17]. This is especially true when, under experimental conditions, the placenta is bypassed [6]. However, we speculate that some of the features of the human FAS may be due, in part, to ethanol-associated restriction of placental transport of essential fetal nutrients. Deprivation of essential substances, such as amino acids, could occur regardless of the nutritional status of the mother, resulting in selective fetal malnutrition.

Conclusion

It has been traditionally held that the direct toxic effects of ethanol or its metabolites are responsible for the pathophysiology of the FAS, as well as any potential injury to the offspring of socially drinking women. The results of our experiments indicate that an additional factor must be taken into consideration: ethanol-induced impairment of placental amino acid transport and resultant selective fetal malnutrition.

Summary

The effect of ethanol infusion upon placental uptake of amino acids was studied in pregnant sheep. Blood ethanol levels of 150–260 mg/dl obliterated the normal uptake of amino acids by the in vivo placenta. However, when human placental villi were incubated in vitro with ethanol at 300 mg/dl, there was no inhibition of uptake of ^{14}C alpha-amino isobutyric acid (AIB). In contrast, 2–20 mM, but not 50 or 200 μM, acetaldehyde significantly inhibited AIB uptake (% inhibition at 90 min: 2 mM, 23.5 ± 9.6%; 5 mM, 33.8 ± 4.7%; 10 mM, 54.6 ± 2.9%; 20 mM, 61.5 ± 7.5%; $p < 0.01$ for all 4 concentrations). When human placental villi were preincubated with 10 μM acetaldehyde, washed, and incubated with ^{14}C-AIB in the absence of acetaldehyde, there was significant residual inhibition of AIB uptake. The data suggest that ethanol-associated placental injury may contribute to the pathophysiology of the fetal alcohol syndrome.

References

1 Third Special Report to the US Congress on Alcohol and Health for the Secretary, Department of Health, Education and Welfare (USDNEW, Rockville 1978).
2 Abel, E.L.; Dintcheff, B.A.: Effects of alcohol exposure on growth and development in rats. J. Pharmac. exp. Ther. *207:* 916–921 (1978).
3 Chang, T.; Lewis, J.; Glazko, A.J.: Effect of ethanol and other alcohols on the transport of amino acids and glucose by everted sacs of rat small intestine. Biochim. biophys. Acta *135:* 1000–1007 (1967).

4   Chernoff, G.F.: The fetal alcohol syndrome in mice: an animal model. Teratology *15:* 223–230 (1977).
5   Clarren, S.K.; Smith, D.W.: The fetal alcohol syndrome. New Engl. J. Med. *298:* 1063–1067 (1978).
6   Fisher, S.E.; Barnicle, M.A.; Steis, B.; Holzman, I.; Van Thiel, D.H.: Effects of acute ethanol exposure upon in vivo leucine uptake and protein synthesis in the fetal rat. Pediat. Res. *15:* 335–339 (1981).
7   Henderson, G.I.; Hoyumpa, A.M.; McClain, C.; Schenker, S.: The effects of chronic and acute alcohol administration on fetal development in the rat. Alcohol. clin. exp. Res. *3:* 99–106 (1979).
8   Henderson, G.I.; Turner, D.; Patwardhan, R.V.; Lumeng, L.; Hoyumpa, A.M.; Schenker, S.: Inhibition of placental valine uptake after acute and chronic maternal ethanol consumption. J. Pharmac. exp. Ther. (in press, 1981).
9   Korsten, M.A.; Matsuzaki, S.; Feinman. L.; Lieber, C.S.: High blood acetaldehyde levels after ethanol administration. New Engl. J. Med. *292:* 386–389 (1975).
10 Kouri, M.; Koivula, T.; Koivusalo, M.: Aldehyde dehydrogenase activity in human placenta. Acta pharmac. tox. *40:* 460–464 (1977).
11 Krasner, J.; Tischler, F.; Yaffe, S.J.: Human placental alcohol dehydrogenase. J. Med. *7:* 323–331 (1976).
12 Kronick, J.B.: Terratogentic effects of ethyl alcohol administered to pregnant mice. Am. J. Obstet. Gynec. *124:* 676–680 (1976).
13 Lloyd-Still, J.D.: Malnutrition and intellectual development. (Publishing Sciences Group, Inc. Littleton 1976).
14 Majchrowicz, E.; Mendelson, J.H.: Blood concentrations of acetaldehyde and ethanol in chronic alcoholics. Science, N.Y. *168:* 1100–1102 (1970).
15 Miller, R.K.; Berndt, W.O.: Mechanisms of transport across the placenta: an in vitro approach. Life Sci. *16:* 7–30 (1975).
16 Rawat, A.K.: Effect of maternal ethanol consumption on fetal and neonatal rat hepatic protein synthesis. Biochem. J. *160:* 653–661 (1976).
17 Rawat, A.K.: Fetal alcohol syndrome metabolic abnormalities. Ohio St. med. J. *74:* 109–111 (1978).
18 Smith, C.H.; Adcock, E.W., III; Teasdale, F.; Meschia, G.; Battaglia, F.C.: Placental amino acid uptake: tissue preparation, kinetics, and preincubation effect. Am. J. Physiol. *224:* 558–564 (1973).
19 Wilson, F.A.; Hoyumpa, A.M.: Ethanol and small intestine transport. Gastroenterology *76:* 388–403 (1979).
20 Wunderlich, S.M.; Baliga, B.S.; Munro, H.N.: Rat placental protein synthesis and peptide hormone secretion in relation to malnutrition from protein deficiency or alcohol administration. J. Nutr. *109:* 1534–1541 (1979).

S.E. Fisher, MD, Director of Pediatric Gastroenterology Research, North Shore University Hospital, 300 Community Drive, Manhasset, NY 11030 (USA)

Alcohol Feeding Alters (^{3}H)Dopamine Uptake into Rat Cortical and Brain Stem Synaptosomes

A. T. Tan, R. Dular, I. R. Innes

Department of Pharmacology and Therapeutics, University of Manitoba Faculty of Medicine, Winnipeg, Canada; Centre de Recherche, Hôpital Louis-H.-Lafontaine, Montréal, Qué., Canada

Introduction

Acute and chronic administration of ethanol have each been shown to affect the turnover of dopamine in substantia nigra and caudate nucleus [1, 8, 11], the dopamine receptor sensitivity in nucleus accumbens [6], corpus striatum, diencephalon and limbic area [7, 13, 17], and the release of dopamine from brain slices [3, 5, 14]. These results suggest the involvement of dopaminergic transmission in various stages of alcohol intoxication. Since dopamine has been demonstrated to have an inhibitory action on cortical and brain stem neurons [9, 10], we have investigated the effect of ethanol on the uptake of (^{3}H)dopamine into nerve terminals (synaptosomes) isolated from two regions of rat brain.

Materials and Methods

Animals

For acute ethanol treatment, ethanol (2 g/kg, as 30% ethanol in saline) was administered intraperitoneally to Long Evans rats weighing 400–450 g. An equal volume of saline was given to control rats. Rats were killed by decapitation 30 min or 16 h after a single injection of ethanol or saline.

Chronic alcoholic rats were prepared by giving either water (control) or ethanol in water as the drinking fluid for Long Evans rats (200 g). Ethanol concentration was gradually increased, being 5, 7.5 and 10% for the 1st, 2nd and 3rd weeks, respectively, and increased to 15% from the 4th to the 8th week. Both groups of rats were given the same food except that the food for control rats was supplemented with dextrose equivalent to the calories provided by ethanol consumption. Chronic alcoholic rats were killed by decapitation when they were still on ethanol treatment (0 time), or after ethanol was replaced by water for 48 or 120 h before the animals were killed.

Chemicals

(^{3}H)dopamine, spec. act. 6.45 Ci/mmol, was obtained from New England Nuclear. All other chemicals were analytical reagent grade commercial products.

Preparation of Synaptosomal Fractions

Brains were removed and cortex and brain stem, including diencephalon, dissected. They were quickly weighed and homogenized in ice-cold 0.32 M sucrose in a glass-Teflon homogenizer. The homogenates were centrifuged at 1,000 g for 10 min and the supernatants were subjected to a further centrifugation for 20 min at 10,000 g. The resulting pellets (P_2) which are rich in nerve endings (synaptosomes) [19] and in which the synaptosomes are responsible for the uptake of dopamine to this fraction [16] were used for uptake studies. The whole preparation was carried out at $-5\,°C$.

(^{3}H)Dopamine Uptake Studies

P_2 fractions were resuspended in buffer with the following composition: NaCl 150 mM, KCl 5 mM, MgSO$_4$ 1.2 mM, Na$_2$HPO$_4$ 1.2 mM, CaCl$_2$ 1.5 mM, Tris 10 mM, and glucose 10 mM, pH 7.4. After 30 min preincubation at room temperature, the P_2 suspensions were pipetted into microcentrifuge tubes which contained (^{3}H)dopamine and were mixed thoroughly by vortexing. The uptake proceeded for 5 min at room temperature and was terminated by centrifugation in a microcentrifuge. The supernatants were removed by aspiration and pellets dissolved in 1% sodium dodecyl sulfate. The radioactivities in the pellets were determined by a Beckman LS8100 liquid scintillation counter with Scintiprep 2 as counting medium. Protein concentrations were estimated by the method of *Lowry* et al. [12] with bovine serum albumin as the standard.

Results

(^{3}H)dopamine uptake into both cortical and brain stem synaptosomes were inhibited in chronic alcoholic rats killed while still under alcohol (fig. 1). In the brain stem, the inhibition persisted although ethanol had been withdrawn for 48 and 120 h. At 120 h after withdrawal, the inhibition appeared to be even more pronounced than that found in rats still under alcohol. In contrast, in the cortex at 48 h withdrawal from alcohol there was a large increase of dopamine uptake. Under our experimental conditions (^{3}H)dopamine uptake into cortical synaptosomes after 48 h withdrawal was almost double that of rats still under alcohol, and was about 40% larger than that of the control. The increase persisted even after 120 h withdrawal from alcohol.

The pattern of dopamine uptake into both cortical and brain stem synaptosomes 30 min after a single dose of ethanol treatment was similar to that of the chronic rats, i.e. there was inhibition of uptake in both areas (fig. 2). However, 16 h after a single dose of ethanol treatment, the uptake

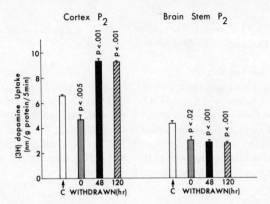

Fig. 1. Effects of chronic ethanol treatment on high affinity (^{3}H)dopamine uptake into rat cortex and brain stem synaptosomal fractions (P_2), in controls (c) □, under ethanol (0) ▨, and at 48 h ■ and 120 h ▨ withdrawal from ethanol. (Values refer to the mean ± SE of 3 experiments. p refers to the level of significant difference from control).

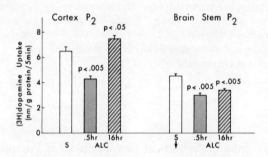

Fig. 2. High affinity (^{3}H)dopamine uptake into rat cortex and brain stem P_2 fractions in controls (s), after a single dose injection of ethanol (2.5 g/kg i.p.) □ at 0.5 h ▨ and at 16 h ▨ withdrawal from the injection. (Values are mean ± SE of 4 experiments, p refers to the level of significant difference from control).

of dopamine into cortical synaptosomes was only slightly larger than that of the control, while brain stem synaptosomes still showed an inhibition compared with the control, although less than the inhibition 30 min after treatment.

The presence of 50 and 100 mM of ethanol in the incubation medium caused a 13 and 17% decrease in dopamine uptake into brain stem synaptosomes but had no effect on uptake into cortical synaptosomes (table I).

Table I. In vitro effect of ethanol on high affinity (^{3}H)dopamine uptake into rat cortical and brain stem P_2 fractions

Treatment	(^{3}H)dopamine uptake, nmol/g protein/5 min	
	cortex P_2	brain stem P_2
Control	7.20 ± 0.54	5.62 ± 0.11
50 mM	7.23 ± 0.34 n.s.	4.89 ± 0.19, p<0.05
100 mM	6.59 ± 0.24 n.s.	4.65 ± 0.20, p<0.01

Values are mean ± SEM of 4 experiments. p refers to the level of significant difference from control. n.s. = Not significant.

Discussion

The data in the present investigation demonstrate the effect of ethanol treatment on the uptake of dopamine into synaptosomes. The changes in uptake vary with specific brain regions studied and with the duration of exposure to ethanol. When rats were still under ethanol, with either acute or chronic treatment, there was inhibition of uptake of dopamine by both cortical and brain stem synaptosomes. Since uptake of dopamine to the nerve endings is generally believed to be responsible for the termination of its transmission action [16], a decrease in uptake would prolong its postsynaptic inhibitory effect, which may explain the antianxiety and hypnotic actions of ethanol [4].

Upon withdrawal from ethanol for a certain period of time, the inhibition of dopamine uptake persisted in the brain stem but disappeared in the cortex. In the acute alcohol rats the uptake of dopamine into cortex 16 h after a single injection of ethanol was only slightly (about 10%) more than that of the control, but in the chronic rats after 48 h withdrawal from ethanol the uptake was 43% greater than control and remained at 41% more than control at 120 h withdrawal. In one experiment with chronic ethanol rats, 16 h after discontinuation of ethanol treatment, the uptake of dopamine to cortical synaptosomes was only about 80% of the control value (data not shown). These results suggest that cortical presynaptic dopamine uptake systems in chronic ethanol-treated rats have undergone a more 'permanent' type of change which was not observed in rats injected with a single dose of ethanol. It is tempting to speculate that the accelerated rate of removal of dopamine by the cortical synaptosomes may be connected with the withdrawal symptoms such as seizures and tremulousness, which are most evident 2–3 days after ethanol withdrawal [15, 18]. This speculation is supported by the report that intracerebral injection of dopamine into mice undergoing withdrawal from ethanol significantly depressed the withdrawal reaction [2].

The *in vitro* effects of ethanol on dopamine uptake to synaptosomes as shown in table I rule out the direct effect of ethanol on the uptake into cortical synaptosomes but may explain, at least in part, the inhibitory effects observed in brain stem synaptosomes. Further experiments are required to clarify the reasons for the changes of uptake into cortical synaptosomes and if indeed ethanol is still present in brain stem synaptosomes to account for the inhibition of uptake in alcohol rats.

Conclusion

Both acute and chronic alcohol treatment inhibited the high affinity uptake of (^{3}H)dopamine into cortical and brain stem synaptosomes by about 30% when rats were still under alcohol. 16 h after acute alcohol treatment, dopamine uptake into cortical synaptosomes was slightly elevated from that of control. In chronic alcohol rats, 48 and 120 h withdrawal from alcohol, the uptake of dopamine were 43 and 41% more than that of the control. The inhibition of dopamine uptake into brain stem synaptosomes persisted even after 120 h of withdrawal. The presence of 50 and 100 mM alcohol in the incubation medium did not alter the uptake of dopamine into cortical but inhibited that of the brain stem synaptosomes by 13 and 17%, respectively. Significant changes of dopamine uptake into synaptosomes under various stages of alcohol intoxication may suggest the effect of alcohol on dopaminergic system in brain.

Summary

The uptake of (^{3}H)dopamine into nerve endings (synaptosomes) isolated from cortex and brain stem of rats under various stages of ethanol intoxication were studied. Acute or chronic ethanol treatment inhibited the high affinity uptake of dopamine into both regions by about 30% when rats were still under ethanol. The inhibition in brain stem persisted even after 120 h of withdrawal from ethanol. In cortex, 16 h after acute treatment, the uptake was slightly (about 10%) elevated from that of control in contrast to chronic rats, which were 43 and 41% more than that of the control after withdrawal from ethanol for 48 and 120 h, respectively.

References

1 Bacopoulos, N.G.; Bhatnager, R.K.; Van Orden, L.S., III: The effect of subhypnotic doses of ethanol on regional catecholamine turnover. J. Pharmac. exp. Ther. *204:* 1–10 (1978).
2 Blum, K.; Enbanks, J.D.; Wallace, J.E.; Schwertner, H.A.: Suppression of ethanol withdrawal by dopamine. Experientia *32:* 493–495 (1976).
3 Carmichael, F.J.; Israel, Y.: Effects of ethanol on neurotransmitter release by rat brain cortical slices. J. Pharmac. exp. Ther. *193:* 824–834 (1975).
4 Cole, J.O.; Davis, J.M.: in Arieti, American handbook of psychiatry, vol. 5, pp. 427–440 (Basic Books, New York 1975).
5 Darden, J.H.; Hunt, W.A.: Reduction of striatal dopamine release during an ethanol withdrawal syndrome. J. Neurochem. *29:* 1143–1145 (1977).
6 Engel, J.; Liljequist, S.: The effect of long-term ethanol treatment on the sensitivity of the dopamine receptors in the nucleus accumbens. Psychopharmacology *49:* 247–253 (1976).
7 Hoffman, P.L.; Tabakoff, B.: Alterations in dopamine receptor sensitivity by chronic ethanol treatment. Nature, Lond. *268:* 551–553 (1977).
8 Hunt, W.A.; Majchrowicz, E.: Alterations in the turnover of brain norepinephrine and dopamine in alcohol-dependent rats. J. Neurochem. *23:* 549–552 (1974).
9 Krnjevic, K.; Phillis, J.W.: Actions of certain amines on cerebral cortical neurones. Br. J. Pharmacol. *20:* 471–490 (1963).
10 Krnjevic, K.: Chemical nature of synaptic transmission in vertebrates. Physiol. Rev. *54:* 418–540 (1974).
11 Lai, H.; Makons, W.L.; Horita, A.; Leung, H.: Effects of ethanol on turnover and function of striatal dopamine. Psychopharmacology *61:* 1–9 (1979).
12 Lowry, O.H.; Rosebrough, N.J.; Farr, A.L.; Randall, R.J.: Protein measurement with the Folin phenol reagent. J. biol. Chem. *193:* 265–275 (1951).
13 Seeber, U.; Kuschins, K.: Effects of ethanol and barbiturates on the sensitivity of dopamine-stimulated adenylate cyclase in the corpus striatum of rat. Arch. Pharmacol. *293:* R-8 (1976).
14 Seeman, P.; Lee, T.: The dopamine-releasing action of neuroleptics and ethanol. J. Pharmac. exp. Ther. *190:* 131–140 (1974).
15 Sellers, E.M.; Kalant, H.: Drug therapy-alcohol intoxication and withdrawal. New Engl. J. Med. *294:* 757–762 (1976).
16 Snyder, S.H.; Kuhar, M.J.; Green, A.I.; Coyle, J.T.; Shaskan, E.G.: Uptake and subcellular localization of neurotransmitters in the brain. Int. Rev. Neurobiol. *13:* 127–158 (1970).
17 Tabakoff, B.; Hoffman, P.L.; Ritzmann, R.F.: Dopamine receptor function after chronic ingestion of ethanol. Life Sci. *23:* 643–648 (1978).

18 Victor, M.: The alcohol withdrawal syndrome. Theory and Practice. Postgrad. Med. *47:* 68–72 (1970).
19 Whittaker, V.P.; Michaelson, I.A.; Kirkland, R.J.A.: The separation of synaptic vesicles from nerve-ending particles (synaptosomes). Biochem. J. *90:* 293–303 (1964).

Dr. A.T. Tan, Centre de Recherche,
Hôpital Louis-H.-Lafontaine, Montreal, PQ HIN 3M5 (Canada)

Subject Index

Acetaldehyde 112, 126, 219
Acetylcholine 49
Adrenocorticotropic hormone 20, 21, 36
Alcohol dehydrogenase 26, 155, 209
Aldehyde dehydrogenase 155, 167, 209
Amino acids 119, 216, 218
Arginine 63, 119

Biogenic amines 50

Catecholamines 16, 46, 86, 162
Choline 49
Circadian rhythm 20
CNS stimulants 162, 169
Contraceptives 131, 135, 175
Corticosterone 100, 103
Corticotropin-releasing factor 20
Cortisol 3
Cushing's syndrome 20, 100
Cyclic AMP 40, 45

Disulfiram 168
Dopamine 19, 224

Endorphins 44
Enkephalin 150
Estradiol 28, 172, 216
Ethanol
   cardiovascular effects 3
   hypothalamus 15
   voluntary drinking 179, 190, 196, 205

Fetus 1, 51, 70, 79, 83, 99, 115, 122
Flutamide 172, 211

GABA 17, 47, 124
Growth hormone 17, 19, 40, 52, 58, 87, 91, 101

Histamine 48, 162
Histidine 118
Homovanillic acid 48, 162
Hypogonadism 24, 130
Hypothalamic hormones 17, 119

Ketoacidosis 78
Korsakoff's syndrome 17

Leydig cells 26
Liver disease 24
Luteinizing hormone 18, 28, 40, 64

Marihuana 143
Miniature dams 179
Miniature sows 190

Naloxone 16

Ornithine decarboxylase 99
Oxytocin 43, 143

Placenta 216
Prolactin 64

Reproduction 25, 75, 101, 143

Tamoxifen 206
Testosterone 26, 39, 145, 147, 175
Δ^9-Tetrahydrocannabinol 145, 149
Thyrotropin-releasing hormone 8, 11, 19, 37

Vasopressin 17, 19, 43

Wilson's disease 30